Exploring Thalamocortical Interactions

Exploring Thalamocortical Interactions

Circuitry for Sensation, Action, and Cognition

W. Martin Usrey

Department of Neurobiology, Physiology and Behavior;
Department of Neurology; and the Center for Neuroscience,
University of California, Davis

S. Murray Sherman

Department of Neurobiology, University of Chicago

OXFORD
UNIVERSITY PRESS

Oxford University Press is a department of the University of Oxford. It furthers
the University's objective of excellence in research, scholarship, and education
by publishing worldwide. Oxford is a registered trade mark of Oxford University
Press in the UK and certain other countries.

Published in the United States of America by Oxford University Press
198 Madison Avenue, New York, NY 10016, United States of America.

Library of Congress Cataloging-in-Publication Data
Names: Usrey, W. Martin, author. | Sherman, S. Murray, author.
Title: Exploring thalamocortical interactions : Circuitry for Sensation, Action,
and Cognition / Usrey, W. Martin, Sherman, S. Murray
Description: New York, NY : Oxford University Press, [2022] | Includes
bibliographical references and index.
Identifiers: LCCN 2021023850 (print) | LCCN 2021023851 (ebook) |
ISBN 9780197503874 (paperback) | ISBN 9780197503898 (epub) |
ISBN 9780197503904 (ebook)
Subjects: MESH: Cerebral Cortex—physiology | Thalamus—physiology |
Neural Pathways—physiology | Synaptic Transmission—physiology
Classification: LCC QP383 (print) | LCC QP383 (ebook) |
NLM WL 307 | DDC 612.8/25—dc23
LC record available at https://lccn.loc.gov/2021023850
LC ebook record available at https://lccn.loc.gov/2021023851

DOI: 10.1093/med/9780197503874.001.0001

9 8 7 6 5 4 3 2 1

Printed by Sheridan Books, Inc., United States of America

We dedicate this book to Ray Guillery (1929–2017), who was a wonderful colleague, role model, and great inspiration to our careers and personal lives.

Contents

Preface

This book can be seen partly as an evolution of ideas on thalamus and cortex expressed in three prior books by one of us (SMS) and Ray Guillery (Sherman and Guillery, 2001, 2006, 2013). We say "partly," because this book involves a different team with WMU as the new partner. This has resulted in new perspectives on issues raised in the previously mentioned books and the addition of new topics not previously explored. We thus believe that this book is distinct.

The main idea of the book is to provide a basis of cell and circuit features of thalamus and cortex derived from a wide range of experimental approaches and provide what we regard as novel interpretations of this evidence. Much of our analysis involves notions of how the process of evolution places certain constraints on our theoretical framework regarding these issues. In doing so, we challenge much conventional thinking about thalamocortical relationships and functioning of cortex.

It is important to emphasize in this context that much of the book involves our speculation regarding the functional significance of known cell and circuit features of thalamus and cortex. We have attempted to clearly distinguish accepted data and interpretations from such speculation. However, our main goal in this regard is not so much to convince the reader that our thoughts and hypotheses are correct or especially worthy as it is to convince the reader that conventional views of the functional organization of thalamocortical relationships need to be reconsidered.

There has been a groundswell of new and renewed interest in the thalamus and thalamocortical interactions over the past decade, and this has led to new conferences and workshops for the broad community of scientists at all career levels to share in their knowledge and enthusiasm for the subject. We have benefited tremendously from these interactions, which have helped shape our thinking and motivate the writing of this book. Because this book is targeted largely to neuroscientists at early career levels, namely, graduate and postdoctoral students, we asked a selection thereof for feedback on an earlier draft of this book, and we are most grateful for their constructive criticisms. These include Mary Baldwin (from the laboratory of Leah Krubitzer), Naomi Bean and Scott Smyre (from the laboratory of Barry Stein), Heather Macomber (from the laboratory of Mark Sheffield), Pantea Moghimi (from the laboratory of David Freedman), and Takuma Sonoda (from the laboratory of Chinfei Chen). Members of the authors' laboratories also provided helpful feedback, including Brianna Carroll and Christina Mo (from the laboratory of SMS) and Alyssa Sanchez (from the laboratory of WMU). We also thank Marjorie Sherman for her heroic proof-reading of the entire manuscript.

Finally, we wish to acknowledge that, during the writing of this book, the authors' laboratories were supported by the following grants from the US government: SMS has been supported by grants from the National Institutes of Health, NS094184, NS113922, and EY022388; WMU has been supported by grants from the National Institutes of Health, EY013588, EY029880, and EY012576 and from the National Science Foundation, 201601841.

Introduction

The purpose of this book is to offer a somewhat different view from standard (read: text-book) accounts of the relationships between the thalamus and cortex and what this all means for cortical functioning. Some of these ideas have been evolving for some time (Usrey, 2002b; Sherman and Guillery, 2006; Sherman and Guillery, 2013; Briggs and Usrey, 2014; Usrey and Alitto, 2015; Sherman, 2016; Halassa and Sherman, 2019; Usrey and Sherman, 2019). This is not meant as a thorough documentation of all things thalamic and cortical, but rather a selective interpretation of certain features of thalamus and cortex that lead to some new ideas and hypotheses. Some of these are quite speculative, and we shall attempt to emphasize differences in our version between generally accepted facts and speculation. Our goal is not so much to get the reader to accept our hypotheses and speculations, but rather to encourage skepticism and rethinking of standard textbook accounts of the subject.

Most of this monograph is focused on sensory processing involving thalamocortical functioning. However, we do offer limited but targeted generalizations of principles gleaned from sensory processing to overall thalamocortical functioning. It is also worth identifying areas important to the thalamus and cortex that are not much covered here. In particular, we do not in any useful detail discuss development, sleep, or clinical correlates.

1.1 General Features of Thalamus and Cortex
1.1.1 Ventral and Dorsal Thalamus

Thalamus proper can be divided into three parts based on embryonic origin: epithalamus, ventral thalamus, and dorsal thalamus (Figure 1.1A). Epithalamus and most of ventral thalamus need not concern us further. Most of thalamus is dorsal thalamus, and this includes all of the thalamic nuclei that contain cells projecting to cortex; these thalamic neurons that project to cortex are generally known as *relay cells*. Ventral thalamus includes the thalamic reticular nucleus, and this is a structure of much importance in thalamocortical relationships. However, the thalamic reticular nucleus contains no relay cells. Rather, its cells all use γ-aminobutyric acid (GABA) and are thus termed GABAergic; these provide inhibitory

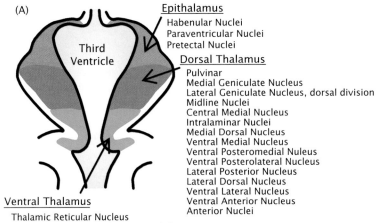

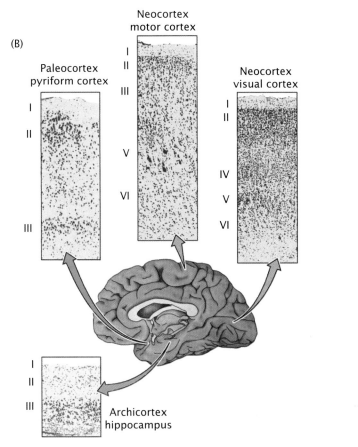

FIGURE 1.1 Different types of thalamus and cortex. A. Thalamus, showing the divisions at an early embryonic stage of a typical mammal. B. Cortex.

input to relay cells of the dorsal thalamus (Smith et al., 1987; Benson et al., 1991; Wang et al., 2001; Campbell et al., 2020). A peculiarity of the ventral thalamus is that it includes the ventral division of the lateral geniculate nucleus, a structure that receives retinal input but does not project to cortex. It is the dorsal division of the lateral geniculate nucleus, which is part of the dorsal thalamus that receives retinal input and projects to cortex. For terminological simplicity in the rest of this account, unless otherwise indicated, we use the term "thalamus" to refer to dorsal thalamus and the term "lateral geniculate nucleus" to refer to its dorsal division.

1.1.2 Cortical Divisions

Figure 1.1B shows the three types of cortex. Paleocortex and archicortex are older in terms of evolution and have fewer than six layers. Paleocortex is the pyriform cortex, which is also the primary olfactory cortex. Archicortex is the hippocampal formation. Neocortex has expanded greatly in mammalian evolution and represents the main cortical subject of this account. Hereafter, unless otherwise indicated, when we refer to "cortex," we mean neocortex.

1.1.3 Overview of Thalamocortical Relationships

Figure 1.2 shows some general features of thalamus and cortex. As general features, these are meant to give a simple overall view of certain relationships, but, as will be shown later in this account, this is only a first approximation of these relationships, and exceptions do exist.

The thalamus is a paired structure near the center of the brain (dashed black outline in medial view of cortex). Each part of the pair in humans is roughly the size of a walnut. The main body of thalamus is comprised of nuclei in which relay cells reside—that is, cells that project to cortex—and thus these nuclei comprise the dorsal thalamus. The color coding indicates roughly the pattern of thalamocortical projections. The thalamic reticular nucleus is also shown. For the most part, it fits snugly against the lateral surface of the thalamus (Figure 1.3) and is shown in Figure 1.2 as cut away and distanced from the thalamus purely for illustrative purposes.

Four further points need to be emphasized. First, connections between thalamus and cortex in both directions are ubiquitous. Every cortical area appears to receive a thalamic input, and often (maybe even always) inputs from multiple thalamic nuclei converge on a single cortical area. Likewise, every thalamic nucleus receives an input from cortex. However, an important detail is that all thalamic nuclei are innervated by layer 6 cells from one or more areas, whereas a subset of these nuclei, in addition, are innervated by layer 5 cells from one or more areas. The difference between these layer 5 and layer 6 corticothalamic projections is significant and is covered more fully in Chapters 5, 6, and 8.

Second, virtually all information reaching cortex must pass through thalamus.[1] For instance, the retina does not directly innervate cortex but instead must pass its information

---

1. A possible exception is the path taken by olfactory information to neocortex (reviewed in Tham et al., 2009; Merrick et al., 2014). Olfactory axons are sent from the olfactory bulb to pyriform (olfactory) cortex, which is paleocortex. Connections from pyriform cortex to neocortex involve two routes: a direct one and one that involves a thalamic relay via the medial dorsal nucleus. We emphasize in Chapter 5 that many pathways in the brain evolved to perform a modulatory rather than an information-bearing function. This raises the possibility, currently

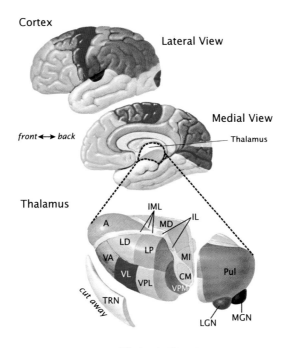

Thalamic Structures

A (Anterior Nuclei)	MI (Midline Nuclei)
CM (Central Medial Nucleus)	MGN (Medial Geniculate Nucleus)
IL (Intralaminar Nuclei)	Pul (Pulvinar)
IML (Internal Medullary Lamina)	TRN (Thalamic Reticular Nucleus)
LD (Lateral dorsal Nucleus)	VA (Ventral Anterior Nucleus)
LGN (Lateral Geniculate Nucleus)	VL (Ventral Lateral Nucleus)
LP (Lateral Posterior Nucleus)	VPl (Ventral Posterolateral Nucleus)
MD (Medial Dorsal Nucleus)	VPm (Ventral Posteromedial Nucleus)

FIGURE 1.2 Thalamocortical relationships shown schematically. At the top are two views of cortex, and the dashed blue outline shows the location of the thalamus at the center of the brain. Below is an enlarged view of the left thalamus, showing the main relay nuclei and the thalamic reticular nucleus. The thalamic reticular nucleus normally sits tight up against the lateral surface of the relay nuclei (see Figure 1.3), but it is cut off and pulled away to reveal the relay nuclei. The color coding shows the main regions of cortex to which each thalamic nucleus projects. Redrawn (courtesy of Mary Baldwin) from Netter, 1977.

on to thalamus for relay to cortex. Why is this? Whatever the reason, this places thalamus at a most strategic position in information flow, because, if relay cells are strongly enough inhibited by GABAergic inputs, local or otherwise, information they receive will not be passed on to cortex. In this sense, the thalamus acts like a gate for information flow to cortex. This would seem to be an efficient way to block unwanted information, because to block an information stream at the cortical level would require silencing orders of magnitude more cells and synapses than to do this at the level of thalamus. Thalamus thus appears to serve as a strategic bottleneck for information flow to cortex. It should be further noted that a gate can

untested, that the direct pathway from pyriform cortex to neocortex is modulatory rather than information-bearing. If so, and if the route through the medial dorsal nucleus is an information-bearing one, then it follows that even olfactory information reaches neocortex via a thalamic relay.

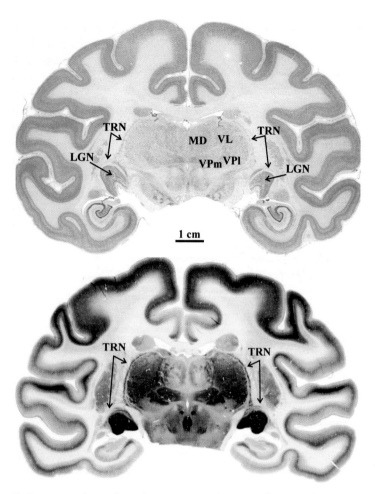

FIGURE 1.3 Thalamic reticular nucleus shown in coronal sections from monkey brain. Notice close apposition of the thalamic reticular nucleus to lateral surface of dorsal thalamus. *Upper*: Nissl stain. *Lower*: Staining for GABAergic neurons.

Abbreviations: *LGN*, lateral geniculate nucleus; *MD*, medial dorsal nucleus; *TRN*, thalamic reticular nucleus; *VL*, ventral lateral nucleus; *VPm/l*, medial and lateral divisions of ventral posterior nucleus.

be completely shut, completely open, or ajar, and likewise, the flow of information through thalamus can be blocked, be a torrent, or be a moderate flow.

Third, gating is not the only function subserved by thalamic relays. Cell and circuit functions can affect the temporal properties of the messages relayed, and these can have important consequences. A prime example of this is a property of thalamic relay cells to fire in different response modes, tonic or burst, which are controlled by voltage-gated Ca^{2+} channels in their somadendritic membranes. This is further explored in Chapter 3. Other more subtle effects of the thalamic relay, including spike[2] timing and attention, are considered in Chapters 10 and 12.

2. We use the terms "spike" and "action potential" throughout interchangeably to mean the same thing. That is, these terms refer to all-or-none regenerative and propagating voltage phenomena that can be either Na^+ or Ca^{2+} based.

Fourth, whereas the flow of information to cortex must typically pass through thalamus, the flow out of cortex to other cortical areas or to subcortical targets does not require a thalamic passage, with one notable exception. That is, there exist cortico-thalamo-cortical circuits, involving layer 5 output cells, and this is a main subject of Chapters 6 and 13.

One final point about the thalamus acting like a relay. "Relay" is often taken to mean a machine-like, often electronic, infallible, and unchanged passage of a message. However, the main definition in the Oxford and Webster dictionaries is to a set of fresh horses substituted for tired ones, as was used in stage coach travel. We regard the thalamic relay as more like a relay inn, where the horses would be changed, the travelers might eat a meal, spend a night on their own or with others, and leave refreshed and perhaps significantly changed. And some may not move on at all from the inn.

1.2 Changing Views of Thalamic Functioning

It seems not too far-fetched to assert that our views of thalamus have advanced from that of a boring relay to our current one that there is a lot of interesting stuff going on there. Much of the earlier disregard of thalamus resulted from the great success of the receptive field approach to the study of sensory processing. This is embodied in the great collaborative work of David Hubel and Torsten Wiesel, work for which they shared the 1981 Nobel Prize with Roger Sperry (Hubel and Wiesel, 1998; Wurtz, 2009). Basically, visual receptive fields become increasingly elaborated as the functional synaptic hierarchy is ascended from retinal receptor through to extrastriate visual cortical areas. This elaboration is thought to underlie how the brain processes the visual world. One exception to this noted by Hubel and Wiesel is the retinogeniculate synapse, across which there seemed to be little or no receptive field elaboration (Hubel and Wiesel, 1961). This observation was repeated often, suggesting little useful function for the lateral geniculate nucleus, but it should be noted that most of these early studies were performed in anesthetized animals.

We now understand that these older observations can be reinterpreted in view of more recent data. That is, the lateral geniculate nucleus has a unique role in visual processing: unlike the rest of synaptic levels that are chiefly involved in receptive field elaboration, the retinogeniculate synapse is a place at which the flow of information is modulated in several important ways not involving major effects on the spatial organization of receptive fields before being sent on to visual cortex. Such modulation includes, among other features, altering the gain of the retinogeniculate synapse and controlling various voltage- and time-dependent ion channels in relay cells that affects temporal properties of their responses. This indeed is a main point of this account, and these properties of the thalamic relay are emphasized in the rest of the monograph.

Another general area of maturation of our appreciation of thalamus is that all of it serves definable functions. Although we generally understood from early neuroanatomical studies that all nuclei of the dorsal thalamus innervate cortex, the specific function of most, until fairly recently, has been poorly appreciated. This is because in many cases we lacked an understanding of what information was being relayed to cortex. For example, we think we understand that the main role of the lateral geniculate nucleus is to relay retinal information to cortex. A similar appreciation has long existed for other sensory thalamic relays and certain motor relays known to pass on cerebellar information. However, for most of thalamus, the information relayed, and thus the main function of the nucleus, was unknown. This led to either ignoring these thalamic nuclei in general theoretical frameworks for thalamocortical processing or assigning them a vague "association" function that was simply a role easily ignored and forgotten. We now believe that we know the information relayed by most or all of these largely ignored thalamic nuclei: this information derives from cortex itself. This point is elaborated in Chapter 6 and is also a main point of this account.

1.3 General Aims of This Book

1.3.1 A Challenge to Conventional Views of Thalamocortical Relationships

Our main goal here is to challenge some basic tenets of thalamocortical relationships as described in textbooks and offer alternative views. The basic account we challenge is shown in Figure 1.4, using the somatosensory system as an example. The interpretation is that information comes in from the periphery (i.e., the primary spinal afferent with a cell body in the dorsal root ganglion), and the information arrives at the thalamic level (i.e., the ventral posterior nucleus) from which it is relayed to primary sensory cortex. From there, the information is sent up a cortical hierarchy through various stages until it reaches motor cortex. From motor cortex, a message is sent subcortically, typically through various brainstem motor sites, to eventually influence motoneurons and thus behavior.

We take issue with several aspects of the scheme in Figure 1.4.

- Any time a new sensory receptor or peripheral sensory process evolves, it will have no survival value if it lacks a fairly immediate motor output. Thus, the idea that an evolutionary process would result in such a lengthy and time-consuming process before a sensory signal can conceivably produce a motor response seems implausible. We review newer evidence in Chapter 13 indicating that multiple steps in sensorimotor processing, including the earliest steps, involve motor outputs.
- An interpretation of Figure 1.4 is based on the idea of following the information as it is processed and then passed on from the periphery through thalamus and cortex and finally to motoneurons. The pathways shown in Figure 1.4 are based largely on anatomical data, and the implicit assumption is that all the pathways shown contribute more or less equally to the transmission of information. We provide evidence in Chapter 5 that this is not the case and that many pathways in and between thalamus and cortex provide modulatory functions rather than transmitting information. To follow the flow of information as it is processed by brain circuits, which is a fundamental early step in understanding information processing by the brain, it is necessary to identify the subset of pathways that do the heavy lifting of information transmission.
- In this scheme, once information reaches cortex, its further processing stays entirely within cortex until the executive level of motor cortex is reached. This means that there is no involvement of thalamus in such processing beyond the initial thalamic relay stage. Among

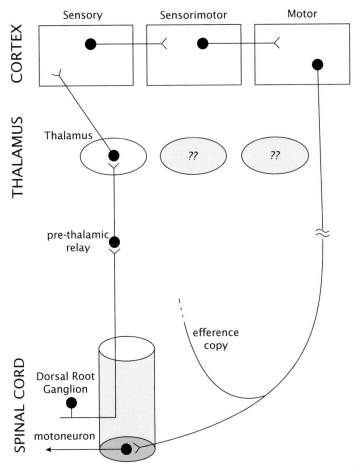

FIGURE 1.4 Prevailing (textbook) view of thalamocortical relationships. See text for details.

other issues, this provides no role for most of thalamus, by volume, which is indicated in Figure 1.4 with question marks. We make the case in Chapter 6 that these thalamic nuclei uninvolved in the information (as depicted in Figure 1.4) are actually very much involved in cortical functioning at all levels of sensorimotor processing.

• While it is true that most textbook accounts do not get into the subject of efference copies (also known as *corollary discharges*), these are actually quite important, because they allow us to disambiguate actual sensory changes caused by events in the environment from sensory changes caused by our own movements. For instance, we normally make 3–5 saccadic eye movements per second, providing a stimulus at the retina of the world rotating each time, but we do not normally experience an unstable visual environment. This is because a copy of the instruction to move the eyes is sent to early stages of visual processing, including thalamic levels, to negate the effects of the impending eye movement. This copy is the *efference copy*. A mark of the importance of efference copies for any organism that moves is that it is a ubiquitous feature of the animal kingdom. The subject of efference copies, when treated at all, is often dealt with in an extremely abstract manner, with little or no neuronal substrate, and, where such a substrate is discussed, it is often vaguely placed near or at the motor end of an extensive sensorimotor chain. This ambiguity regarding most textbook accounts of efference copies, if such accounts are even included, is indicated in Figure 1.4 by a branch of an efferent axon heading off to an undefined target. We argue in Chapter 13, quite speculatively, that efference copies are an important part of thalamo-cortical processing and that the complexity of circuitry involving efference copies has been barely investigated.

1.3.2 Scope and Overall Plan for the Remaining Chapters

The organization of the following chapters is as follows:

- Chapters 2–4 provide basic cell and circuit features of thalamus and cortex to provide a basis for all readers to digest the content of the following chapters that challenge conventional wisdom. These chapters also provide a basis for understanding the properties that underlie various thalamic relay functions, such as gating and the burst and tonic firing modes of relay cells.
- Chapter 5 introduces the concept that glutamatergic pathways in thalamus and cortex are not homogeneous but rather can be classified as *driver* or *modulator* types. The hypothesis here is that only the driver subset is responsible for the bulk of information transfer between neurons, and thus identifying these inputs is a necessary prerequisite for further understanding of information processing in the brain.
- Chapter 6 makes the case that, based on its driver input, thalamic relays can be divided into *first order* and *higher order*. First order relays receive driver input from a subcortical source (e.g., retinal input to the lateral geniculate nucleus), whereas higher order relays receive theirs from layer 5 of a cortical region (e.g., much of pulvinar, which is a higher order thalamic relay), thus participating in transthalamic, cortico-thalamo-cortical pathways.
- Chapter 7 goes into further details of thalamic circuitry, and Chapter 8 does this for cortical circuitry.
- Chapter 9 notes the heterogeneity of thalamocortical and corticothalamic circuits and discusses the importance of obtaining a complete classification of these. This underscores the breadth of different functions supported by these connections between thalamus and cortex.
- Chapter 10 explores more detailed physiological features of thalamic relay properties with an emphasis on the role of spike timing in improving information exchange between neurons.
- Chapter 11 presents evidence for parallel processing in thalamocortical and corticothalamic circuits.
- Chapter 12 describes certain behavioral effects on thalamocortical functioning, such as the effects of alertness and attention on circuit functioning, including effects on burst and tonic firing of relay cells. It also offers suggestions regarding the possible evolutionary implications of attention as regards cortical processing.
- Chapter 13, which is particularly speculative, explores the links between thalamocortical processing and efference copies.

- Finally, Chapter 14 attempts a summing up and synthesis of the ideas put forward throughout this book.

We also feel it important to emphasize important gaps in our knowledge. Thus, we end each of the subsequent chapters with a list of key questions that we believe are particularly important to recognize and answer.

2

Cell Types in the Thalamus and Cortex

Broadly speaking, the dorsal thalamus is comprised of two major cell types: *relay cells* that project to the cerebral cortex and *local interneurons* (Figure 2.1A and B). An additional cell type with origins in the ventral thalamus, *cells of the thalamic reticular nucleus*, must also be included in the discussion of thalamocortical interactions as they integrate excitatory signals from the dorsal thalamus and cortex and, in turn, provide inhibitory input to thalamic relay cells (Figure 2.1C). Finally, a small number of thalamic cells, primarily in the intralaminar and midline nuclei, project to targets in the striatum and amygdala. Although these cells are not discussed further in this chapter, they will be considered in Chapter 9.

Neurons in all layers of the cerebral cortex receive thalamic input via their basal and/or apical dendrites; however, this input is not equal to all cortical neurons. Thalamic axons arborize most extensively in cortical layer 4, followed by the overlying and deeper layers. The cortex also sends excitatory projections to the thalamus; layer 6 neurons provide modulatory input to all thalamic neurons, whereas layer 5 neurons provide driving input to higher order thalamic neurons. A review of the distinctions between driving and modulatory inputs is presented in Chapter 5. In the following sections, we examine the general properties of thalamic and cortical cells that are integral to thalamocortical interactions.

2.1 Thalamus

The two major cell types in core divisions[1] of the dorsal thalamus, relay cells and interneurons, are further subdivided and categorized into distinct groups based on their anatomy, neurochemistry, and physiology. Although early investigations often labeled cells based on

1. A suggested division for thalamocortical projections is into "core" and "matrix" moieties (Jones, 1998), which is discussed more fully in Chapter 9. Briefly, core projections are highly topographic and mainly innervate the middle cortical layers, especially layer 4, whereas matrix projections are rather diffuse and target upper layers, especially layer 1. This chapter is focused on relay cells located in core divisions of the thalamus.

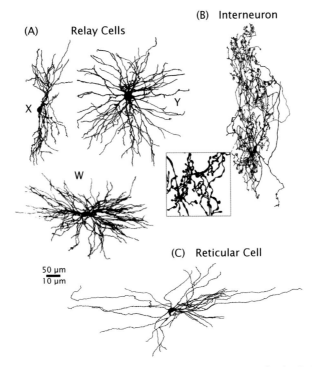

FIGURE 2.1 Cell types in the cat's thalamus. Camera lucida reconstructions of individual thalamic neurons labeled with intracellular injection of horseradish peroxidase (HRP). A. Examples of three types of relay neurons in the lateral geniculate nucleus (X, Y, and W types). Each cell type receives input from a different class of retinal ganglion cells and gives rise to specific projection patterns to primary visual cortex. Typical of these cell types, the dendritic arbor of the X cell is tufted and is oriented perpendicular to the plane of the layers, the W cell arbor is oriented parallel to the layers, whereas the arbor of the Y cell is more spherical. B. A GABAergic interneuron from the lateral geniculate nucleus. Compared to the relay cells, the cell body of the interneuron is smaller and the dendrites extend perpendicular to the layering for a greater distance, generally spanning the entire width of the layer. The inset shows a magnified view of a portion of the dendritic arbor. C. A GABAergic neuron from the thalamic reticular nucleus. The dendrites of the reticular cell are thinner and less dense than those of the relay cells. Data from Friedlander et al., 1979; Stanford et al., 1981, 1983; Hamos et al., 1985; Uhlrich et al., 1991.

the single feature under examination, it is now clear that there are distinct categories/classes of cells based on their shared characteristics. In Chapters 7 and 11, we consider species-specific properties of thalamic neurons in the context of evolution and homology. Here we focus on the general distinctions of thalamic relay cells and interneurons.

2.1.1 Relay Cells

2.1.1.1 General Properties

Thalamic relay cells are critical for providing the cortex with information about the external environment. Relay cells also provide input to the thalamic reticular nucleus via branches of their thalamocortical axons (Figure 2.2). In both structures, relay cells excite target neurons via the release of the neurotransmitter, glutamate. Similar to the cortical neurons that they innervate, the driving input onto thalamic cells is subserved by a minority of their synaptic inputs.

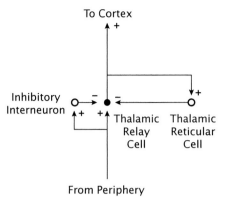

FIGURE 2.2 Schematic diagram of the connections between thalamic relay cells, interneurons, and reticular cells. Relay cells receive feedforward excitatory input from the periphery, feedforward inhibition from local interneurons, and feedback inhibition from thalamic reticular cells.

For instance, in the visual system where retinal ganglion cells are the source of driving input to neurons in the lateral geniculate nucleus, only about 5–10% of the synapses made onto lateral geniculate neurons are from retinal ganglion cells (Van Horn et al., 2000). In contrast, feedback axons from the cortex provide approximately 30–50% of the synapses with lateral geniculate relay cells. The remaining inputs come from the brainstem, local interneurons, and neurons in the thalamic reticular nucleus. Similar values are reported for neurons in the ventral posterior nucleus that receive feedforward input from lemniscal axons carrying somatosensory information and in the medial geniculate nucleus that receive feedforward auditory input from the inferior colliculus (reviewed in Sherman and Guillery, 2013). Although feedforward inputs comprise just a small percentage of the synapses made onto relay cells, this input is sufficient to establish the general structure of thalamic receptive fields (Hubel and Wiesel, 1961; Cleland et al., 1971a,b; Usrey et al., 1999). The other sources of input to relay cells play a more limited role in shaping thalamic receptive fields, but instead play a prominent role in determining whether and how relay cells respond to their feedforward input. The importance of this latter role cannot be overstated, as the thalamus is the gateway to the cortex, and the temporal structure of signals sent to cortex is a major determinant in the success of thalamocortical communication (Usrey et al., 2000; Swadlow and Gusev, 2001; Bruno and Sakmann, 2006).

2.1.1.2 Classification: Anatomy, Neurochemistry, and Physiology

Relay cells in core regions of the sensory thalamus have been classified using many criteria, and these classification schemes facilitate the discussion and understanding of the roles they serve in cortical processing. In general, relay cell axons terminate densely and in a topographically restricted fashion in cortical layer 4. Some relay cells also give rise to a collateral projection that targets cortical layer 6, and yet other relay cells have axons that eschew layer 4 and arborize in layers 2/3. The functions served by these different projection patterns has been the subject of much investigation, and, in many cases, these projections align along cell classification schemes that distinguish parallel processing streams (discussed below in Chapter 9).

In the somatosensory, auditory, and visual pathways, relay cells with axons targeting layer 4 are located in the ventral posterior nucleus, the ventral division of the medial geniculate nucleus, and the principal layers of the lateral geniculate nucleus, respectively (e.g., the parvocellular and magnocellular layers in primates, the A layers in carnivores, and the core region in rodents; Figure 2.3). A separate population of lateral geniculate relay cells (e.g., the koniocellular layers in primates, the C layers in carnivores, and the shell region in rodents) have axons that avoid layer 4 and project, instead, to the overlying supragranular layers (layers 2 and 3; Fitzpatrick et al., 1983; Lachica and Casagrande, 1992; Usrey et al., 1992). Furthermore, projections from the somatosensory and auditory thalamus to layers 2/3 also exist in primary somatosensory and auditory cortices (Viaene et al., 2011b).

Within the layer 4 and layers 2/3 projection pathways to cortex there exist additional distinctions in relay cell types that form the basis of multiple parallel streams of information flow to cortex. These parallel streams are particularly evident in the visual system, where relay cells can be sorted on the basis of their ocular input (ipsilateral, contralateral, binocular) and their responses to increases and decreases in the brightness of a centered stimulus (on cells, off cells, on/off cells) (reviewed in Usrey and Alitto, 2015). These cell types are mapped further using additional classification schemes that distinguish cells on the basis

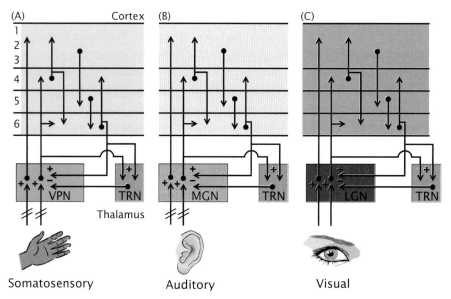

FIGURE 2.3 Circuit diagram illustrating similarities in the relationship between first order thalamic neurons, neurons in the thalamic reticular nucleus, and neurons in the cerebral cortex. For the somatosensory (A), auditory (B), and visual (C) pathways, relay cells send sensory signals mainly to cortical layers 4 and 6 or to layers 2/3 (note: experimental evidence for the layers 2/3 projection is more complete for the visual pathway than for the somatosensory or auditory pathways). These signals are processed by networks of cortical neurons (a general motif of connections is illustrated for excitatory cells from different layers). Layer 6 corticothalamic neurons complete a reciprocal loop of information exchange and provide excitatory feedback to the relay cells, as well as onto neighboring interneurons and cells in the thalamic reticular nucleus that then inhibit relay cell activity.

Abbreviations: *LGN*, lateral geniculate nucleus; *MGN*, medial geniculate nucleus; *TRN*, thalamic reticular nucleus; *VPN*, ventral posterior nucleus. Adapted from Alitto and Usrey, 2003.

of receptive field size, morphological features, response latency, sensitivity to contrast, time course of response, extent of linear summation of visual responses, and selectivity for chromatic stimuli, to name a few (reviewed in Usrey and Alitto, 2015). Collectively, these efforts distinguish three major cell types in both cats and monkeys: the X, Y, and W cells (Figure 2.1A) and the parvocellular, magnocellular, and koniocellular cells, respectively. While the specific details of these cell types and the parallel retino-geniculo-cortical streams to which they belong are the focus of Chapter 11, here we wish to emphasize that each stream (e.g., ipsi/on/X cells, contra/off/Y cells, etc.) contains a complete map of visual space, and their retinal ganglion cell counterparts completely tile the retina. Thus, there are at least eight separate and complete representations of visual space in the thalamocortical projections of geniculate X and Y relay cells, and the number of representations is likely more than 20 when the diversity of W cells and their projections are included. Once these segregated signals arrive in primary visual cortex, different cortical circuits then integrate their information to varying extents (e.g., binocular cells with overlapping on/off receptive fields) before sending them on to downstream targets for further processing.

Compared to relay cells projecting to layer 4, those projecting to layers 2/3 typically have smaller and thinner cell bodies, axons, and synaptic terminals, and they appear lighter with certain anatomical stains (e.g., thionin or cresyl violet) (Figure 2.1; Guillery, 1970). These morphological distinctions are accompanied by physiological differences. In particular, the conduction velocities for action potentials are slower for layers 2/3 projecting cells than for layer 4 projecting cells (Irvin et al., 1986). Because relay cells targeting the disynaptic pathway to layers 2/3 have faster conduction velocities than relay cells providing monosynaptic input to layers 2/3, signals traveling in the three pathways to layers 2/3 could arrive at approximately the same time, thus positioning layers 2/3 neurons as potential coincidence detectors. While the idea of coincidence detection between the pathways is attractive, more experiments are needed to determine whether and how signals traveling in more than one pathway interact to produce postsynaptic responses.

2.1.2 Interneurons

2.1.2.1 General Properties

All thalamic nuclei contain GABAergic interneurons. Although interneurons are outnumbered by relay cells (by 3–5 to 1, depending on the species and nucleus[2]), they play a prominent role in governing the activity of relay cells and in augmenting the effects of lateral inhibition that are typically established in upstream structures (e.g., the surround of center/surround receptive fields). Similar to relay cells, interneurons receive a mix of feedforward and feedback input in addition to modulatory inputs from the brainstem (reviewed in

---

2. Exceptions seem to be rats and mice, which have interneurons in their lateral geniculate nucleus, but few if any are found in other thalamic nuclei (Arcelli et al., 1997). Because triads and glomeruli seem related to interneuronal dendritic terminals, these structures are also rare in these animals outside of the lateral geniculate nucleus. Other mammals so far studied, including other rodents, generally have interneurons, triads, and glomeruli throughout thalamus. However, this point has been questioned by evidence that the lateral posterior nucleus of the rat has a substantial fraction of interneurons (Li et al., 2003b).

Jones, 2007). The receptive fields of interneurons are also generally similar to those of relay cells (Sherman and Friedlander, 1988; Hirsch et al., 2015). With these similarities in mind, interneurons are distinct from relay cells in other respects. Compared to relay cells, the cell bodies and dendrites of interneurons are smaller, thinner, and more extensively branched, respectively (Figure 2.1B). Interneuron dendrites also often extend further from their cell bodies than those of relay cells. As a result of these anatomical features, interneurons are electrotonically far less compact than are relay cells (Bloomfield and Sherman, 1989). Finally, whereas interneurons have axons with conventional synaptic terminals, their dendrites also contain synaptic terminals that participate in triadic synaptic circuits (see below and Chapters 4 and 7). These differences in dendritic morphology are accompanied by differences in synaptic organization, as discussed below.

2.1.2.2 Types of Synapses

Similar to other neurons, interneurons have axons that make synapses with target cells. In addition to these axodendritic synapses, thalamic interneurons are capable of releasing neurotransmitter from their dendrites via dendrodendritic synapses (Colonnier and Guillery, 1964; Sherman and Guillery, 1996; Cox and Sherman, 2000; Sherman and Guillery, 2013). Because the more distal regions of interneuron dendrites are likely to be electrotonically isolated from the spike-generating axon hillock, interneurons may be able to multitask, with their dendrodendritic synapses acting independently of their axodendritic synapses (Sherman, 2004). Although this feature of interneuron physiology is discussed in greater depth in Chapters 4 and 7, it is worth mentioning here that dendrodendritic synapses participate in the formation of triads, a unique organization of synapses between the processes of three cell types that contact each other in extremely close proximity. Within a triad, there is a feedforward excitatory synapse shared with the dendrites of a relay cell and a neighboring interneuron. The interneuron, in turn, forms a dendrodendritic inhibitory synapse onto the relay cell's dendrite adjacent to the excitatory input. As might be imagined, this circuit could serve many purposes, including adjusting the gain (i.e., responsiveness) of relay cell activity.

2.1.3 Cells of the Thalamic Reticular Nucleus

Cells in the thalamic reticular nucleus receive synaptic input from the collaterals of relay cell axons targeting the cerebral cortex and from the collaterals of corticothalamic feedback axons that originate from layer 6 neurons and target thalamic relay cells and interneurons (Guillery and Harting, 2003; Pinault, 2004; Usrey and Alitto, 2015). Recent evidence indicates that an additional input to the thalamic reticular nucleus emanates from layer 5 of some cortical areas (Prasad et al., 2020). Because thalamic reticular cells are all GABAergic and have axons that target the same thalamic nuclei from which they receive input, thalamic reticular cells are in a strategic position to integrate ongoing thalamic and cortical activity in order to modulate thalamocortical communication. Along these lines, it is noteworthy that thalamic reticular cells also receive input from a variety of sources thought to be involved with arousal and attention, including the brainstem reticular formation and basal forebrain (Lee and Dan, 2012; Beierlein, 2014). Consequently, thalamic reticular cells, via their input to relay cells, may play a role in attentional modulation of relay cell activity and in thalamocortical communication (discussed in Chapter 12). Separate from functions for sensory processing, thalamic reticular cells also play a major role in establishing thalamic activity patterns associated with sleep. Although not discussed further here, several reviews on sleep and activity in the thalamic reticular nucleus are available for further reading (Lee and Dan, 2012; Halassa and Acsady, 2016; Vantomme et al., 2019).

2.2 Cerebral Cortex

Similar to the thalamus, the cerebral cortex is comprised of projection neurons and interneurons. The projection neurons are all pyramidal cells (Figure 2.4) and use glutamate for synaptic transmission. They also have axon collaterals that provide synaptic input to other neurons in the same cortical area. In contrast, cortical interneurons can be excitatory or inhibitory, using glutamate or GABA, respectively, as a neurotransmitter (Figure 2.4) and, as their name implies, have axons remain exclusively within the given cortical area. As a general rule, excitatory neurons (both projection neurons and interneurons) have spiny dendrites (Figure 2.5), whereas inhibitory neurons (all interneurons) have smooth dendrites. This distinction has made it possible to infer cellular function on basis of morphology. As described further below in this chapter and in Chapter 8, this inference has limitations, as the functions performed by cortical neurons are far more complex than can be predicted simply on the nature of their neurotransmitter,[3] depending critically on the organization of their inputs and outputs within the cortical architecture, which includes columns and layers. A general working model of this cortical architecture is shown in Figure 2.3. Relevant to the topics in this chapter, thalamic axons make synapses primarily in layer 4, but also in all other layers; layers 2/3 neurons provide feedforward and feedback input to other cortical areas; layer 5 neurons provide feedforward input to higher order thalamic nuclei; and neurons in layer 6 provide feedback input to first order and higher order thalamic nuclei.

2.2.1 Layer 4 Neurons

Layer 4 neurons that receive thalamic input include glutamatergic spiny neurons and GABAergic smooth neurons. In some animals, including tree shrews and a variety of primates, the spiny neurons in layer 4 lack apical dendrites (i.e., they are nonpyramidal stellate cells; Figure 2.4B), whereas in others, including rodents and carnivores, layer 4 spiny neurons are mostly, if not exclusively, pyramidal cells. With just a few exceptions, layer 4 neurons are generally all interneurons; thus, they receive feedforward signals, integrate these signals with other sources of input, and provide output only to cells within the local area of cortex, typically to neurons in overlying layers 2 and 3 and, to a lesser extent, layer 6 (Figure 2.3).

Similar to relay cells in the thalamus, a minority of the synapses made with layer 4 neurons come from feedforward inputs (Peters and Payne, 1993; Peters et al., 1994). For neurons in primary sensory areas of the cortex, this feedforward input comes from thalamic

---

3. In Chapter 5, it is noted that glutamate can produce an inhibitory postsynaptic effect if it activates Group II metabotropic receptors.

(A)

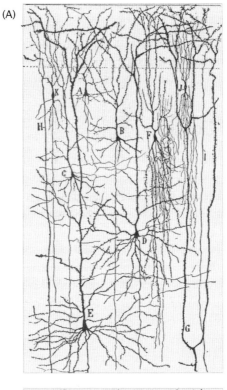

(B)

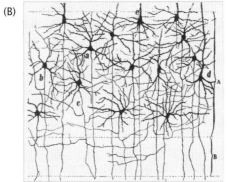

FIGURE 2.4 Drawings of cortical neurons from Santiago Ramon y Cajal. A. Examples of cell types from the superficial layers of the human precentral gyrus. This figure nicely illustrates the morphology of small (*A, B,* and *C*) and medium (*D* and *E*) pyramidal cells. Additional cell types shown are bitufted cells (F and J) and a fusiform cell (*K*). Also included are the dendritic shafts of deeper pyramidal cells (*G, H,* and *I*). B. Examples of stellate cells from the visual region of the cat cortex. In this figure, *A* and *B* indicate the layer of stellate cells and layer of giant pyramidal cells, respectively; *a, b,* and *c* correspond to stellate cells with descending axons, and *d* and *e* correspond to intermingled pyramidal cells. Modified from Cajal, 1899a, 1899b.

(A)   (B)

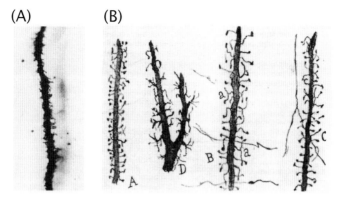

FIGURE 2.5 Preparations from Santiago Ramon y Cajal showing the spiny dendrites of cortical neurons. A. Photomicrograph showing the dendritic spines on the apical dendrite of layer 5 neurons from cat visual cortex. B. Four drawings showing spines of pyramidal cells from the rabbit (*A*), child (*B*), cat (*C and D*). Modified from Cajal, 1933; DeFelipe and Jones, 1988.

relay cells; in higher cortical areas, this input comes from higher order thalamic nuclei and other cortical areas (see Chapters 5, 7, and 9 for a discussion on the potential driver/modulator functions of these pathways). Interestingly, thalamic input to layer 4 is not exclusive for layer 4 neurons. For instance, in visual cortex of the cat, only 65% of lateral geniculate synapses are made onto layer 4 cells; the remainder are made onto the apical dendrites of cells in layers 5 and 6 (Peters and Payne, 1993). Moreover, because layer 4 neurons receive convergent input from multiple lateral geniculate cells (e.g., ~30 lateral geniculate cells in the cat provide an average of 5 synapses each onto a given layer 4 cell[4]; Reid and Usrey, 2004), the effectiveness of any given lateral geniculate cell in driving a postsynaptic response is relatively weak (Reid and Alonso, 1995). Consequently, effective thalamocortical communication depends on the coordinated activity of multiple inputs (Usrey et al., 2000; Usrey, 2002a,b; Bruno and Sakmann, 2006), which underlies the emergence of receptive fields that are dramatically different from those of their individual afferents, as in the case of binocular-, orientation-, and often direction-selective simple cells in cat visual cortex (Reid and Alonso, 1995; Alonso et al., 2001). Although we explore these topics more completely in Chapters 8 and 10, the important take-home message for now is that layer 4 cells integrate input from a variety of sources to establish new receptive fields and response properties that are not necessarily evident in their individual presynaptic inputs.

2.2.2 Layer 5 Neurons

Layer 5 pyramidal cells come in two flavors. One resides mostly in the upper half of layer 5 (often known as layer 5a) and the other in the lower half (or layer 5b).

4. On average, individual layer 4 cells in the cat have approximately 2,500 synapses, of which about 150 (6%) come from the axons of geniculate relay cells (Peters and Payne, 1993).

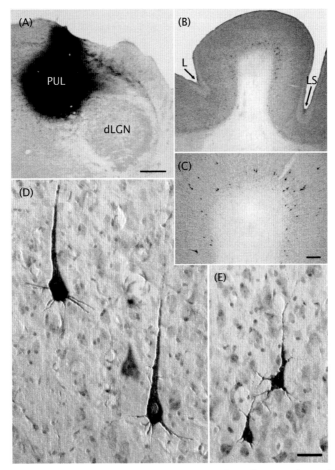

FIGURE 2.6 Corticothalamic cells in the cat with projections to the pulvinar nucleus, a higher order tha-lamic nucleus. A. Following the injection of horseradish peroxidase into the pulvinar nucleus (A), cortical areas 5 and 7 have retrograde labeled cells in layers 5 and 6 (B and C). D and E. Higher-power views of retrograde labeled cells. Corticopulvinar cells in layer 5 (D) are larger than those in layer 6 (E). Scale bars: 1 mm in A,B; 250 μm in C; and 30 μm in D,E. Modified from Baldauf et al., 2005.

2.2.2.1 Layer 5a Pyramidal Cells

These cells are relatively small pyramidal cells, with apical dendrites that typically fall short of reaching layer 1. They do not have subcortical projections but are the source of interareal corticocortical projections (Llano and Sherman, 2009; Petrof et al., 2012). They typically re-spond to excitatory inputs with single spikes rather than bursts (Llano and Sherman, 2009).

2.2.2.2 Layer 5b Pyramidal Cells

These are the largest neurons in cortex and typically have apical dendrites that reach layer 1 and arborize there (Figure 2.6). They also give rise to prominent subcortical projections (Petrof et al., 2012), targeting higher order thalamic nuclei, the basal ganglia, numerous brainstem targets, and, in some cases, the spinal cord (Bourassa and Deschênes, 1995;

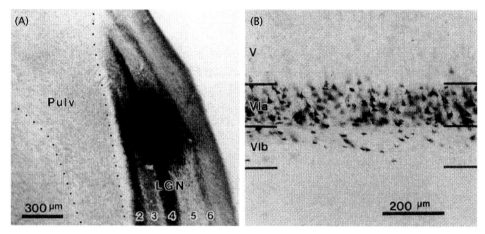

FIGURE 2.7 Distribution of corticogeniculate neurons in the tree shrew. A. A wheat germ agglutinin-horseradish peroxidase (WGA-HRP) injection site restricted to the lateral geniculate nucleus. To illustrate the layers in the geniculate, an injection of WGA-HRP was also made in the contralateral eye. LGN: lateral geniculate nucleus; Pulv: pulvinar nucleus. B. The resulting pattern of labeled neurons in visual cortex. Cell bodies are located across the depth of layer 6 with the density greatest in layer 6A. Modified from Usrey and Fitzpatrick, 1996.

Levesque et al., 1996; Veinante et al., 2000; Kita and Kita, 2012; Prasad et al., 2020). Often these multiple targets are innervated by single, branching axons; these are discussed more fully in Chapter 13. The only other prominent subcortical projections emanate from layer 6 cells that provide input largely limited to thalamus or claustrum.[5] It should be emphasized that the layer 5 subcortical projections to motor centers in the brainstem and spinal cord represent the only known route by which cortex can fairly directly influence behavior. This point is taken up in more depth in Chapters 12 and 13.

These layer 5b pyramidal cells can fire in either tonic or burst mode, the latter due to the activation of a voltage-gated conductance via Ca^{2+} channels located in the dendrites (Larkum et al., 1999; Llano and Sherman, 2009). The Ca^{2+} conductance seems to play an interesting role in coincidence detection: that is, an excitatory input to basal dendrites alone evokes only tonic firing, but when paired with an excitatory input arriving at the apical tufts in layer 1, a Ca^{2+} spike is evoked, leading to burst firing (Larkum et al., 1999; Llano and Sherman, 2009).

2.2.3 Layer 6 Neurons

Layer 6 has a special relationship with the thalamus. Thalamic axons targeting layer 4 frequently branch to innervate neurons in layer 6,[6] and layer 6 contains pyramidal cells with feedback projections to the thalamus (Figure 2.7; Jones, 2007; Sherman and Guillery, 2013).

---

5. A population of layer 6 cells, distinct from corticothalamic cells, projects to the claustrum, which, in turn, projects back to the cortex.

6. Although thalamic branching in layer 6 appears widespread, it remains to be determined if it is universal across species and cortical areas.

As noted above, other layer 6 cells innervate the claustrum and yet others, that do not project subcortically, innervate other cortical areas (Petrof et al., 2012). Layer 6[7] corticothalamic feedback cells are all glutamatergic, and they provide modulatory input to neurons in both the first order and/or higher order thalamic nuclei that supply their cortical area (Deschênes et al., 1998; Sherman and Guillery, 2013). Corticothalamic cells also have local axon collaterals that innervate neurons in the overlying cortical layers that receive input from the thalamic neurons targeted with their descending axons (Lund and Boothe, 1975). For instance, corticothalamic neurons in primary visual cortex make synapses with relay cells and interneurons in the lateral geniculate nucleus and related reticular thalamic cells as well as with the layer 4 neurons that receive input from the lateral geniculate nucleus (Figure 2.3; Usrey and Fitzpatrick, 1996). Based on these connections, layer 6 corticothalamic neurons are in a strategic position to influence thalamocortical communication both at its thalamic sources as well as at its main cortical target. As will be discussed in Chapters 8 and 11, corticothalamic pathways can include additional complexity, including parallel streams of feedback projections that have a special relationship with parallel feedforward thalamocortical pathways (Briggs and Usrey, 2009).

2.3 Summary and Concluding Remarks

Relay cells and interneurons are the two major cell types in the dorsal thalamus. Relay cells are all glutamatergic and send axons that branch en route to the cortex to provide collateral input to neurons in the thalamic reticular nucleus. Across thalamic nuclei, there are two types of relay cell projections to cortex, one that arborizes extensively in layer 4 and another that bypasses layer 4 to arborize extensively in layers 2 and 3. These two classes of relay cells involved are also typically distinguished from each other in important ways, including their afferent input, stimulus preference, neurochemistry, and axon conduction velocity. Within the layer 4 and layers 2/3 projecting classes of relay cells, there are additional cell classes; thus, the projections from thalamus to cortex are comprised of parallel processing streams, a topic that is discussed in greater depth in Chapter 11. Local interneurons in the thalamus are all inhibitory and release GABA from their axon terminals. Most or all of these local interneurons also release GABA from their dendrites (further discussed in Chapter 7) in a manner that appears to be largely independent of their axonal release.

Within cortex there are also projection neurons (pyramidal) and interneurons (nonpyramidal). Like thalamus, projection neurons (neurons with axons that project subcortically or to another cortical area) are all glutamatergic. In contrast, local interneurons include one group that releases glutamate and another that releases GABA. The GABAergic interneurons include multiple subclasses with distinct morphological and physiological properties. Cortical projections to thalamus arise from neurons in layers 5 and 6. Layer 5 corticothalamic neurons provide strong, driving input to higher order thalamic nuclei,

---

7. To be distinguished from some layer 5 cells that also project to some thalamic nuclei with a very different function from these layer 6 cells, as described in Chapter 6.

whereas layer 6 corticothalamic neurons provide weaker, modulatory input to both higher order and first order thalamic nuclei. These projections terminate on relay cells and interneurons in the thalamus. A collateral branch also provides input to neurons in the thalamic reticular nucleus. Thus, neurons in the thalamic reticular nucleus, like local thalamic interneurons, integrate feedforward and feedback input to influence relay cell activity. As with the relay cell pathways to cortex that include multiple parallel streams for information flow, evidence indicates that at least some of the pathways from cortex to thalamus also contain multiple streams (further discussed in Chapter 11); however, more work is needed to determine whether this is a general property of corticothalamic projections.

2.4 Some Outstanding Questions

1. Do all thalamic pathways to cortex involve parallel pathways, as is the case with the visual system (i.e., W, X, and Y or K, P, and M), and how do the three pathways interact during cortical processing?

2. What role(s) do synaptic triads in the thalamus play in sensory processing?

3. In some areas of the brain, action potentials that travel down branching axons can produce synchronous responses in target neurons. Do thalamic axons that innervate layer 6 and layer 4 neurons serve to synchronize responses of neurons in the two layers, and, if so, how does this synchronous activity affect cortical processing?

4. Some cortical neurons have an apical dendrite, while others lack an apical dendrite. What role does the apical dendrite serve in cortical processing? Is there a general function for the apical dendrite, or does the function vary depending on the laminar position of the cell?

5. Historically, all thalamic neurons with axons that target cortex are termed "relay cells." More recent work indicates that some of these cells provide driving input to cortical neurons, whereas others provide modulatory input. With this in mind, should we reconsider the definition of relay cells? The term "relay" implies the "handing off" or transfer of a signal.

3

Intrinsic Membrane Properties

Each thalamic and cortical neuron receives many synaptic inputs, up to 1,000 or so synapses on the former and 10,000 or so on the latter. A key to understanding the functioning of neural circuits is figuring out how these multiple inputs act in aggregate to affect the firing of the target cell. The vast majority of inputs synapse on dendrites, with rare synapses found on the soma. Dendritic arbors tend to be complex, branching structures (see, e.g., Figure 2.1), further complicating the problem. An active synapse effectively injects an ionic current, positive for excitatory synapses or negative for inhibitory ones, into the cell at the dendritic point of contact (see Chapter 4 for details). The problem to be solved is being able to follow the current as it flows through the branched dendritic arbor toward the soma to depolarize or hyperpolarize the axon hillock (or initial segment of the neuron's axon), which, in turn, controls the firing of the cell. This current can be affected by other active synapses along its route, by details of branches it must traverse, and by many other factors.

3.1 Cable Modeling

The first efforts to address the complicated issue of how current spreads within neurons involved simplifying matters by treating dendritic segments as passive cables. This approach, known as "cable modeling," provides an estimate of the effects of synaptic input at a specified dendritic site on voltage changes at the soma, which, in turn, approximates the voltage changes at the axon hillock and nearby initial axonal segment (Rall, 1969a,b; Jack et al., 1975). Action potential initiation occurs here, and so voltage changes at the hillock determine the firing properties of the neuron.

There are obvious limitations to this approach. Chief among them is the requirement that the dendrites be passive, meaning that they do not support active processes such as voltage- or ligand-dependent conductances and that voltage propagation is strictly electrotonic. Also, various parameters, which cannot usually be measured in neurons, such as electrical properties of the cytoplasm and membrane, are often estimated from measurements made using non-neural cells and assumed to apply to neurons (Jack et al., 1975).

Dendrites resemble electrical cables in the following ways. Like cables, they have an inner core, namely cytoplasm that has low electrical resistance, and an outer covering that has high resistance. If the covering had infinite resistance, any current injected into the cable would flow in both directions through the internal medium without decrease (Figure 3.1A).

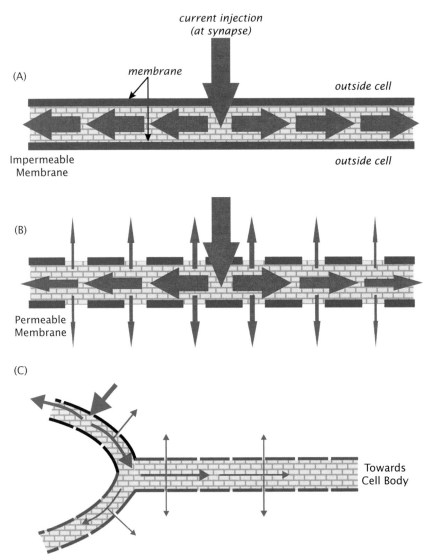

FIGURE 3.1 Schematic view of current flow in dendrites as a result of current injected into the process (e.g., by synaptic activation). The red arrows indicate the direction of current flow, and their width indicates current amplitude. A. Example of an impermeable membrane. All of the current flows within the process, although it divides to flow in both directions from the point of current injection. B. Example with partly permeable membrane. As current flows inside the dendrite, some of it leaks across the membrane, so that the farther from the injection site, the smaller the amplitude of the remaining intracellular current flow. C. Effect of dendritic branching. As the current flow reaches a dendritic branch point, the current divides to enter all of the resultant branches.

Redrawn from Sherman and Guillery, 2013.

However, neuronal membranes are electrically leaky because the membrane contains numerous ion channels that are open at any one time, and these allow current to flow across the membrane in the form of charged ions. Thus, current flowing through the cytoplasm will decrease in amplitude as it travels from the injection site (Figure 3.1B). This current flow is known as passive or *electrotonic*. To complicate things further, as the current flows electronically toward the soma and encounters a dendritic branch point, the current divides, with one part flowing distally and the other toward the soma (Figure 3.1C). More generally, cable modeling can provide a first estimate of synaptic impact, and this is based chiefly on the morphology of the postsynaptic neuron. It is beyond the scope of this account to provide further details of cable modeling; such details can be found elsewhere (Rall, 1969a, b; Jack et al., 1975; Jack et al., 1983; Segev, 1998; Bedard and Destexhe, 2016).

There are a few notable conclusions that can be drawn from cable modeling. One is that, all other parameters being equal, the more distally located a synapse is within the dendritic arbor, the less impact it will have on the target neuron's firing. However, as seen below, this is not always the case. Dendritic segments can be thought of as cylinders. The larger the cylinder's diameter, the larger its cross-sectional area and surface area, but the surface area is only linearly proportional to the diameter, whereas the cross-sectional area increases with the square of the diameter. It therefore follows that the ratio of cross-sectional area to membrane area increases for dendrites with increasing thickness. Cytoplasm has a much lower electrical resistance than does the membrane, and so the larger the ratio of cytoplasmic cross-sectional area to membrane area, which occurs in larger diameter dendrites, allows more current (e.g., from synaptic activation) to flow through the cytoplasm to the soma with less leaking out through the membrane.

3.2 Ionic Conductances

A full description of ionic membrane conductances is beyond the scope of this account. Instead, we summarize those we deem most critical to the functioning of thalamic and cortical neurons. A fuller account of these conductances can be found elsewhere (Huguenard and McCormick, 1994; Spruston et al., 2014).

3.2.1 Passive Conductances

At rest, neurons typically maintain a voltage difference such that the inside of the neuron is roughly −65 to −75 mV negative with respect to the outside. Figure 3.2 illustrates how this is achieved. Imagine a jar containing a solution of various ions, and two compartments are created by bisecting the inside of the jar with an impermeable membrane, such as a rubber wall (Figure 3.2A). The ions cannot cross between compartments. There is an equal balance of positive or negative ions in each compartment, and thus there is no voltage difference between them.

However, neuronal membranes are not completely impermeable to ions because they contain channels and pumps through which various ions can move between compartments; think of the channels as holes in the membrane that allow ions to diffuse down an

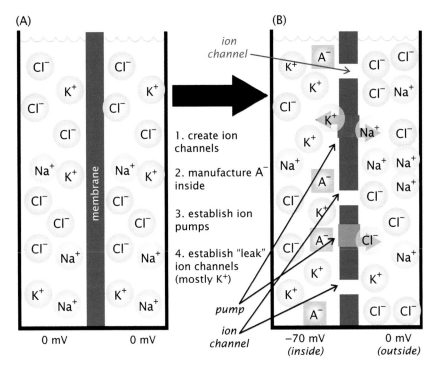

FIGURE 3.2 Hypothetical establishment of membrane potential. A. To start, imagine various ions in solution in a jar with a membrane dividing the contents into two equal chambers. The membrane is impervious to the flow of ions (imagine a rubber sheet), and equal concentrations of the various ions exist in both compartments. There is no voltage difference between the chambers. B. To create the appropriate voltage difference, four factors must ensue as shown (see text for details).

electrochemical gradient and pumps as devices that move ions against an electrochemical gradient (Figure 3.2B). The energy-dependent sodium/potassium (Na^+/K^+) pump transfers Na^+ ions out of the cell and K^+ into the cell (red pump in Figure 3.2B), such that the concentrations of Na^+ and K^+ are greater on the outside and inside of the cell, respectively. With the addition of leak channels inserted into the membranes, Na^+ and K^+ can diffuse down these concentration gradients toward their initial values; however, because leak channels are more permeable to K^+ than to Na^+, more K^+ diffuses across the membrane than Na^+. With this imbalance in charge movement, the inside of the cell develops a negative voltage. K^+ ions will flow out of the cell down their concentration gradient until the concentration gradient is balanced by an opposing electrical gradient established with the outflow of K^+ (the K^+ equilibrium potential). In practice, the resting membrane potential is always less negative than the K^+ equilibrium potential because some Na^+ diffusion occurs through the leak channels. Finally, two additional factors need to be included to approximate the state of a typical neuron: (1) Cl^- pumps that move Cl^- from the inside of the cell to the outside (green pump in Figure 3.2B) and (2) negatively charged proteins, which are thus anions (shown as pink squares labeled A^- in Figure 1.2B) and nucleic acids that are restricted to the inside of cells.

These reversal potentials (also known as *equilibrium potentials*), or E for a particular ion, can be calculated for mammalian neurons from the *Nernst equation*:

$$E_{ion} = \frac{RT}{zF} * In \frac{[ion]_{outside}}{[ion]_{inside}}$$

where R is the gas constant, T is the temperature (in °K), z is the ion charge, F is Faraday's constant, and [ion] is the concentration for the particular ion. The value of E for an ion varies somewhat due mainly to some variation in ionic concentration among neurons. For the three main ions of interest, the approximate values of E are −90 to −100 mV for K^+, −70 to −80 mV for Cl^-, and +55 to +65 mV for Na^+. If only K^+ leaked across the membrane, the resting potential would be at the reversal potential for K^+. The smaller leakage of Na^+ and Cl^-, which are driven by more positive reversal potentials, combine with the K^+ "leak" conductance to create the resting membrane potentials observed, typically between −65 and −75 mV.

3.2.2 Voltage- and Time-Dependent Conductances

Many ion channels are voltage- and time-dependent, meaning that their opening or closing is controlled by membrane potential. The time-dependent factor is attributed to the fact that a change in membrane potential that could affect channel opening takes a certain amount of time to actuate. For example, some K^+ channels are voltage-dependent such that, at resting membrane potentials, they are closed, but sufficient depolarization maintained for a sufficient time will open them. This causes K^+ to flow out of the cell, thereby producing hyperpolarization as positive ions move out of the cell. The action potential, which involves the action of voltage- and time-dependent Na^+ and K^+ channels, is the best-known product of ion movement through voltage-dependent channels. Here, we discuss these channels that are found in the axon and occasionally also in the cell body and dendrites.

3.2.2.1 The Action Potential

Figure 3.3A illustrates the properties of the voltage- and time-dependent Na^+ and K^+ channels that underlie the action potential (Hodgkin and Huxley, 1952; McCormick, 2014). The Na^+ channel has two voltage-sensitive gates, an activation gate and an inactivation gate (Figure 3.3Ai). When the activation gate is open the channel and current (I_{Na}) are *activated*; when closed, these are *deactivated*. When the inactivation gate is open the channel and I_{Na} are *de-inactivated*; when closed, these are *inactivated*. Both gates must be open to allow Na^+ to flow down its electrochemical gradient into the cell and depolarize it. The K^+ channel has only an activation gate, and so the channel and its current (I_K) is *activated* when the gate is open and *deactivated* when the channel is closed. When the gate is open, K^+ flows out of and hyperpolarizes the cell (Figure 3.3Ai). At rest (roughly −65 mV; Figure 3.3Ai), whereas the inactivation gate for Na^+ is open (i.e., the channel is de-inactivated), the activation gate is closed (i.e., the channel is deactivated) and so there is no inward flow of Na^+ ions (I_{Na}). For the K^+ channel, the activation gate is closed, and there is no outward flow of K^+ ions (I_K). With sufficient depolarization (e.g., from an excitatory postsynaptic potential [EPSP]), the Na^+ activation gate opens leading to a large inward I_{Na}, producing the upstroke of the action potential (Figure 3.3Aii and Figure 3.3Av). This depolarization closes the inactivation gate in the Na^+ channel and opens the activation gate in the K^+ channel (Figure 3.3Aiii), thereby causing repolarization of the neuron to its resting level, but the Na^+ inactivation requires about 1 msec,

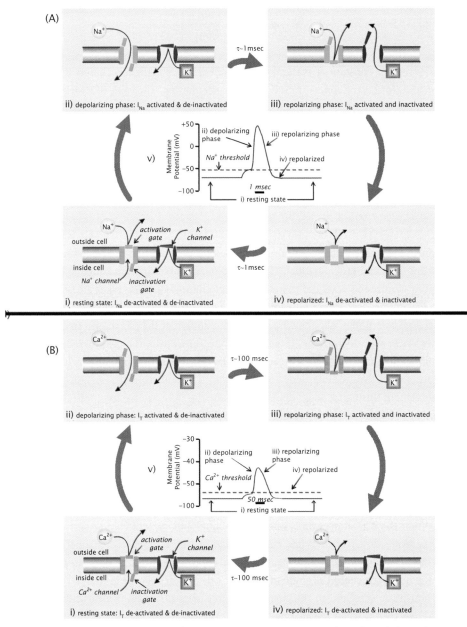

FIGURE 3.3 Schematic representation of voltage- and time-dependent ion channels underlying the conventional action potential and the low threshold Ca²⁺ spike. A. For the action potential, i– iv show the channel events, and v shows the effects on membrane potential. The Na⁺ channel has two voltage-dependent gates: an *activation gate* that opens at depolarized levels and closes at hyperpolarized levels, and an *inactivation gate* with the opposite voltage dependency. Both must be open for the inward, depolarizing Na⁺ current (I_{Na}) to flow. The K⁺ channel (actually an imaginary combination of several different K⁺ channels) has a single activation gate, and when it opens at depolarized levels, an outward, hyperpolarizing K⁺ current is activated. (i) At the resting membrane potential, the activation gate of the Na⁺ channel is closed, and so it is de-activated, but the inactivation gate is open, and so it is also de-inactivated. The single gate for the K⁺ channel is closed, and so the K⁺ channel is also de-activated. (ii) With sufficient depolarization to reach its threshold, the activation gate of the Na⁺ channel opens, allowing Na⁺ to flow into the cell. This depolarizes the cell, leading to the upswing of the action potential. (iii) The inactivation gate of the Na⁺ channel closes after the depolarization is sustained for roughly 1 msec ("roughly," because inactivation is a complex function of time and voltage), and the slower K⁺ channel also opens. These combined channel actions lead to the repolarization of the cell. While the inactivation gate of the Na⁺ channel is closed, the channel is said to

FIGURE 3.3 Continued

be inactivated. (iv) Even though the initial resting potential is reached, the Na⁺ channel remains inactivated because it takes roughly 1 msec ("roughly" having the same meaning as above) of hyperpolarization for de-inactivation. (v) Membrane voltage changes showing action potential corresponding to the events in i–iv). B. For the representation of actions of voltage-and time-dependent T (Ca²⁺) and K⁺ channels underlying the low threshold Ca²⁺ spike, the conventions are as in A; (i–iv) show the channel events, and (v) shows the effects on membrane potential. Note the strong qualitative similarity between the behavior of the T-type Ca²⁺ channel here and the Na⁺ channel shown in A, including the presence of both activation and inactivation gates with similar relative voltage dependencies. (i) At a membrane potential more hyperpolarized than the normal resting potential, the activation gate of the T-type Ca²⁺ channel is closed, but the inactivation gate is open, and so the channel is both de-activated and de-inactivated. The K⁺ channel is also de-activated. (ii) With sufficient depolarization to reach its threshold, the activation gate of the T-type Ca²⁺ channel opens, allowing Ca²⁺ to flow into the cell. This depolarizes the cell, providing the upswing of the low threshold spike. (iii) The inactivation gate of the T-type Ca²⁺ channel closes after roughly 100 msec ("roughly," because, as for the Na⁺ channel in A, closing of the channel is a complex function of time and voltage), inactivating the T-type Ca²⁺ channel, and the K⁺ channel also opens. (iv) These combined actions repolarize the cell. Redrawn from Sherman and Guillery, 2013.

because there is a time as well as a voltage dependency for the process. Although now at the resting membrane potential (Figure 3.3Aiv), approximately 1 msec is needed for the Na⁺ inactivation gate to resume its resting state, which is the open position (Figure 3.3Ai).

Notice that depolarization does not inactivate I_K because the channel has no inactivation gate. However, the inactivation of the Na⁺ channel means that prolonged sufficient depolarization will block I_{Na}, leading to a perhaps counterintuitive result that such depolarization will prevent action potentials rather than being an excitatory event. Also, note that the 1 msec required by the Na⁺ channel to de-inactivate means that action potentials cannot be generated during this time. This is known as the *refractory period*, which at roughly 1 msec limits the firing rate of a neuron to about 1,000 spikes/sec. Because no action potentials are possible during this period, it is often referred to as the *absolute refractory period*.

Typically, K⁺ channels remain open for a brief period following the action potential (not shown in Figure 3.3A), which hyperpolarizes the cell beyond its resting potential. Excitatory inputs during this prolonged hyperpolarization will produce reduced and sometimes no firing. In contrast to the absolute refractory period just described, this period of prolonged K + opening is referred to as the *relative refractory period*.

Figure 3.4A shows the voltage dependency for activation and inactivation of the Na⁺ channel. The ordinate measures the relative I_{Na} flowing through the Na⁺ channels, and the abscissa shows the membrane potential. It should be noted that the curves of Figure 3.4 are drawn from many sources and should be regarded as idealized approximations. These measures are typically made by intracellular recording from a neuron under voltage control (known as *voltage clamp*[1]) by which transmembrane currents are measured in response

---

1. Voltage clamp is a recording technique during which current is passed into or out of the cell via the recording electrode to clamp the membrane voltage to a set value. The current passed to do so reflects the amount needed to offset any ionic currents evoked by a procedure such as a synaptic event, or, as in Figure 3.4 by a sudden voltage shift that would open or close ion channels. A problem is that, to be reliable, this technique requires establishing a uniform transmembrane voltage from the recording site to any ion channels under study. For measuring the curves in Figure 3.4A for the Na⁺ channels underlying the action potential, the relevant Na⁺ channels are in the axon hillock, which is basically at the same point electronically as the recording electrode. However, the Ca²⁺ channels of interest in Figure 3.4B are in the dendrites, overall more distant electronically than are the relevant Na⁺ channels of Figure 3.4A, and so the voltage clamp in this case is less reliable. Thus Figure 3.4B should be treated as a less reliable approximation.

(A)

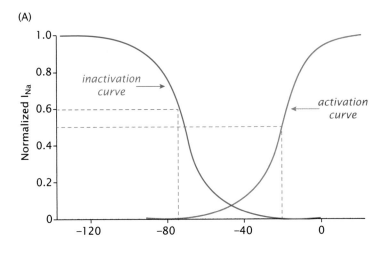

(B)

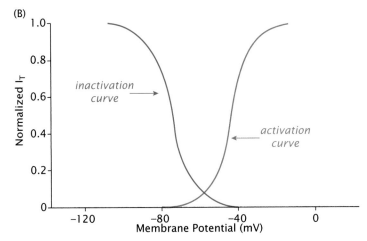

FIGURE 3.4 Activation and inactivation curves for the Na+ and T-type Ca²⁺ channels. These are ideal-ized approximations, and the actual values vary among neurons. Curves such as these are derived via intracellular voltage clamp recording (see text for details). Inactivation curves (red) are created by first holding the neuron at a hyperpolarized potential indicated on the abscissa and stepping to a depolar-ized level that would activate all de-inactivated channels; the resultant relative current for each of these steps is plotted on the ordinate. Activation curves (blue) are created by first holding the neuron at such a hyperpolarized level that all relevant channels would be de-inactivated and then stepping up to the membrane potentials shown on the abscissa; again, the resultant relative current for each of these steps is plotted on the ordinate. A. Activation and inactivation curves for the Na+ channel. B. Activation and inactivation curves for the T-type Ca²⁺ channel. Details in text, including interpretation of dashed lines in A.

to rapid voltage steps. The inactivation curve (red) is created by starting at variable hyperpolarized levels as indicated on the abscissa and stepping up to a very depolarized level such as +20 mV, which completely activates any de-inactivated Na$^+$ channels. By stepping from −120 mV, the maximum I_{Na} is evoked, because all Na$^+$ channels are de-inactivated at −120 mV. The dashed red lines show an example with the starting voltage at −75 mV: a jump to +20 mV evokes a normalized I_{Na} of 0.6. Thus, from less (more) hyperpolarized starting voltages, less (more) I_{Na} is evoked because more (fewer) Na$^+$ channels remain inactivated. One way to read this is that, at various initial membrane voltages, the more hyperpolarized (up to about −120 mV), the more Na$^+$ channels are de-inactivated. The activation curve (blue) is similarly created, except the starting voltage is very hyperpolarized such as −120 mV, thereby de-inactivating all Na$^+$ channels and stepping up to voltages shown on the abscissa. The blue dashed lines show an example for this process: a voltage step from −120 mV to −20 mV evokes a normalized I_{Na} of 0.52.

Typically, the action potential starts at the axon hillock or initial segment and propagates down the axon to its synaptic terminals. Figure 3.5 shows some of the key elements in this process. Once an action potential is evoked at a voltage-dependent Na$^+$ channel, the depolarization will spread electronically to the next voltage-dependent Na$^+$ channel. If the density of such channels is too low (Figure 3.5Ai), the depolarization will be too small to activate the next voltage-dependent Na$^+$ channel, and there is no propagation of the action potential. However, if the channel density is high enough (Figure 3.5Aii), each activated Na$^+$ channel will activate its neighbor, and the action potential thus regenerated travels down the axon.

The refractory period plays an important role in this propagation (Figure 3.5B). Recall that, after repolarization of the action potential, the Na$^+$ channel remains inactivated for about 1 msec. Thus, the traveling action potential leaves a wake of such inactivated channels, meaning that, as it travels down the axon, it cannot activate an action potential traveling in the opposite direction. Even for the slowest conducting action potential, say at 1 m/sec, the wake of inactivated channels would be 1 mm long (and faster traveling action potentials would leave a corresponding longer wake), meaning that electrotonic flow of the excitation effectively cannot spread past the inactivated Na$^+$ channels to activate further ones. All of this ensures that an action potential generated at the axon hillock will travel exclusively orthodromically. Note that under nonphysiological conditions, an antidromic direction for action potential propagation is possible if, for instance, electrical stimulation is applied to the axon at a distance from the soma, but this is an artificial event thought not to occur naturally.

Finally, as bodies and brains evolve to become larger, it becomes more imperative to speed up the propagation of action potentials. One answer to this is the evolution of a *myelin sheath* enwrapping the axon (Figure 3.5C). The myelin sheath prevents ion flow across it, but there are gaps along the sheath that expose the axon membrane: these gaps are called *nodes of Ranvier*. Now the action potential propagates not by activating immediately adjacent membrane, but rather by jumping to the next node of Ranvier where there is a high density of voltage-dependent Na$^+$ channels (Figure 3.5C). This is often known as *saltatory conduction* and provides for much faster propagation. Note also that the refractory period still serves to prevent propagation of the action potential backward.

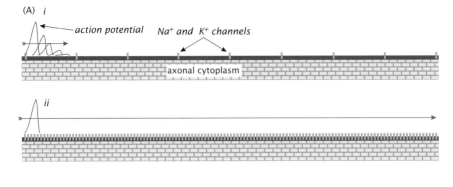

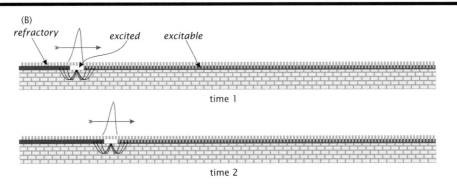

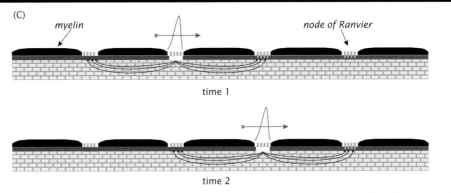

FIGURE 3.5 Schematic views of propagation of the action potential along an axon. A. Effect of Na$^+$ channel density on propagation of action potentials. (i) Low concentration of Na$^+$ (and K$^+$) channels. Current from the depolarizing voltage spike at the left electronically flows to the second Na$^+$ channel, but that channel is so far away the depolarization is subthreshold for activating the second Na$^+$ channel. Thus, the action potential fails to propagate. (ii) High concentration of Na$^+$ (and K$^+$) channels. Now the action potential generated at the left propagates down the axon, because the close packing of Na$^+$ channels means that as each is activated, it sufficiently depolarizes its neighboring Na$^+$ channel to activate it. B. Unmyelinated axon. Current from the action potential leads spreads in both directions inside the axon, depolarizing neighboring patches of axon membrane. In the direction the spike is traveling, the newly depolarized membrane patch will lead to opening of Na$^+$ channels in its forward path, leading to a regeneration of the action potential. This continues in this direction, and so the spike sweeps down the axon. However, the current flow in the reverse direction depolarizes a patch of membrane that recently produced an action potential, and so it is in a refractory period and thus cannot regenerate another one. As a result, the action potential will not reverse direction. The patch of inactivated membrane in the wake of an action potential is 1–50 or more mm in length, which is far too long an extent for current from the activated action potential to span via electrotonic flow. C. Myelinated axon. Myelin acts like an effective insulator, preventing ion flow across it, so current can only flow into and out of the cell at gaps in the myelin where Na$^+$ and K$^+$ channels are especially concentrated; the gaps are known as *nodes of Ranvier*. Thus, instead of a smooth flow of the action potential, as in B, it jumps from node to node. This is saltatory conduction and is much faster than the smooth conduction in B. As in B, the refractory period ensures that the action potential flows in only one direction. B and C redrawn from Sherman and Guillery, 2013.

3.2.2.2 T-type Ca²⁺ Channels

There are a variety of voltage- and time-dependent Ca^{2+} channels found in neurons, but one particularly important for thalamic neurons is the T-type ("T" for "transient") Ca^{2+} channel (Jahnsen and Llinás, 1984a,b; Huguenard and McCormick, 1994; Zhan et al., 1999; Sherman, 2001; Llinás and Steriade, 2006). The voltage gating of this channel is qualitatively quite similar to that of the Na^{+} channel, with both activation and inactivation gates, and, as is the case with Na^{+}, both gates must be open to allow Ca^{2+} to flow down its electrochemical gradient into the cell and depolarize it. However, there are important quantitative differences between these channel properties (Figure 3.3). The terms "activation," "de-activation," "inactivation," and "de-inactivation" have the same meanings for the Ca^{2+} channel and I_{Ca} as described above for the Na^{+} channel and I_{Na}. At rest (roughly −75 mV in this case; Figure 3.3Bi), whereas the inactivation gate for Ca^{2+} is open (i.e., the channel is de-inactivated), the activation gate is closed (i.e., the channel is deactivated), and so there is no inward flow of Ca^{2+} ions (I_T). For K^{+}, the activation gate is closed, and there is no outward flow of I_K. With sufficient depolarization (e.g., from an EPSP; Figure 3.3Bii), the Ca^{2+} activation gate opens, leading to a large inward I_T and producing the depolarizing upstroke (see Figure 3.3Bv). This depolarization inactivates the Ca^{2+} channel and activates the K^{+} channel (Figure 3.3Biii), thereby acting to repolarize the neuron to its resting level, but de-inactivation (i.e., opening the inactivation gate) requires about 100 msec because, as for the Na^{+} channel, there is a time as well as a voltage dependency for the gate changes. Although now at the resting membrane potential (Figure 3.3Biv), again roughly 100 msec at the repolarized level is needed for the gates to resume their resting state (Figure 3.3Bi). This last point means that, like the Na^{+} channel, the Ca^{2+} channel also has an absolute refractory period.

Just as is the case with Na^{+} channels, a sufficiently dense distribution of these T-type Ca^{2+} channels can support a propagating spike. These Ca^{2+} channels are quite ubiquitous in neurons, but, with few exceptions, their density is too low to underlie spike propagation and so they generally just provide for local electrotonic depolarization. However, thalamic neurons are an exception, with dense enough T-type Ca^{2+} channels in their dendrites to propagate spikes.

Figure 3.4B shows the inactivation and de-inactivation curves of the T-type Ca^{2+} channels, derived in much the same way as the curves for the Na^{+} channels in Figure 3.4A.

It should be clear from a comparison of Figure 3.3A and 3.3B that there is great similarity between the voltage-dependent Na^{+} and Ca^{2+} channels, but there are four differences that are especially important:

1. Na^{+} channels are found chiefly in axons, but T-type Ca^{2+} channels are found strictly in somadendritic membranes and not in axons. Thus, spikes from these Ca^{2+} channels can be propagated through the dendritic arbor and soma, but they play no role in propagation of an action potential along the axon.

2. The inactivation time constants are much longer for the Ca^{2+} channels (roughly 100 msec versus 1 msec for the Na^{+} channels), and so it takes a much longer period of depolarization to inactivate these channels and a much longer period of hyperpolarization to de-inactivate them. It should be noted that the short time constants for Na^{+} inactivation

are unusual: most voltage-dependent channels that inactivate have time constants for this much like the T-type Ca²⁺ channel. Also, the inactivation/de-inactivation property for these Ca²⁺ channels is a complex function of voltage and time (Zhan et al., 1999; Smith et al., 2000), such that the more depolarized (hyperpolarized), the faster the channel inactivates (de-inactivates).

3. The extent of depolarization created by a Ca²⁺ spike is smaller than that evoked by a Na⁺ action potential (compare Figure 3.3Av with Figure 3.3Bv).

4. As Figure 3.4 shows, the Ca²⁺ channel operates in a somewhat more hyperpolarized regime than does the Na⁺ channel, and thus thresholds for the various activation and inactivation states are roughly 10–20 mV more hyperpolarized than for the Na⁺ channel. This means that I_T can be activated at a lower threshold than can an action potential, and thus the evoked Ca²⁺ spike is often referred to as the "low-threshold spike."

Regarding this last point, a glance at Figure 3.4 shows that, at a resting potential of, say, −60 mV, the Ca²⁺ channels would effectively be inactivated, and a large enough EPSP would directly activate action potentials. However, if the cell were hyperpolarized enough in time and amplitude to de-inactivate many T-type Ca²⁺ channels by, say, inhibitory synaptic input, the same EPSP would activate a Ca²⁺ spike, which, in turn, would activate action potentials. However, EPSPs are much smaller than Ca²⁺ spikes. Thus the large Ca²⁺ spike typically evokes a high-frequency burst of 2–6 or more action potentials, and this is known as *burst firing*, but when the T-type Ca²⁺ channels are inactivated, each EPSP alone activates fewer action potentials (typically just one) at a lower frequency, and this is known as *tonic firing*. In Chapter 10, we go into more detail regarding the significance of burst and tonic firing.

The relatively long time constant of roughly 100 msec for changing the state of inactivation with voltage changes has two notable consequences. First, the longer refractory period means that the highest firing rate for these Ca²⁺ spikes is about 10 Hz. Second, to de-inactivate the channel requires that it be held sufficiently hyperpolarized for 100 msec or more. We normally expect such hyperpolarization to result from inhibitory synaptic input, but typical inhibitory postsynaptic potentials (IPSPs) seem too brief to accomplish this, although temporal summation of a barrage of IPSPs could extend the time of hyperpolarization sufficiently to de-inactivate these Ca²⁺ channels. Likewise, an EPSP or conventional action potential seems too brief to produce inactivation, although, again, temporal summation could accomplish this. Below, we describe synaptic properties involving activation of metabotropic receptors that produce sufficiently long polarization changes to control effectively the inactivation state of these channels.

3.2.2.3 Other Voltage-Dependent Ca²⁺ Conductances

Several other Ca²⁺ conductances can be activated by depolarization, like the T-type channels, but they differ in having much higher activation thresholds, typically requiring depolarization to about −20 mV. These other types are known as L, N, R, and P/Q channels, and they are located in the dendrites and synaptic terminals (Llinás, 1988; Johnston et al., 1996; Larkum et al., 1999, 2007; Catterall, 2011). They are the reason that Ca^{2+} enters the cell as a result of an action potential. In synaptic terminals, these conductances represent a key link between the action potential and transmitter release, since the action potential will activate these channels, and the resultant Ca^{2+} entry is needed for transmitter release. Less is known about these Ca^{2+} conductances in dendrites, but by providing a regenerative spike that can travel between the site of an activating EPSP and the soma, they may help ensure that distal dendritic inputs that are strong enough to activate this conductance will significantly influence the soma and axon hillock (Larkum et al., 1999, 2007; Suzuki and Larkum, 2020).

These channels often play a role in the dendritic responses to an action potential generated at the axon hillock, responses that come under the rubric of "back propagation" (Stuart et al., 1997). An action potential will depolarize the soma and dendrites, even if passive, via electrotonic current flow, and this, of course, would affect voltage-dependent channels there. This back propagation can be active in cases where a sufficient density of Na^+ or high-threshold Ca^{2+} channels exist in the dendrites, as occurs in some neurons.

3.2.3 Other Conductances

Two other types of ionic conductances are present in neurons. One is ligand-dependent, which is a common property of synaptic transmission and is dealt with in more detail in Chapter 4. This occurs when an external molecule (such as transmitter released from a synaptic terminal) binds to a specific receptor protein in the membrane, the binding leading to opening or closing of various ion channels. The other, which commonly relates to various K^+ channels, is dependent on the internal concentration of Ca^{2+}. In this case, when Ca^{2+} enters the cell (e.g., via opening of voltage-dependent Ca^{2+} channels), certain K^+ channels open, leading to outward flow of K^+.

3.3 Intrinsic Properties of Thalamic Neurons

3.3.1 Cable Properties

We are unaware of any cable modeling of thalamic reticular neurons. However, such cable modeling of thalamic relay cells and interneurons suggests a dramatic difference in how these neurons integrate synaptic inputs. Figure 3.6 illustrates this by showing cable modeling for two relay cells and two interneurons from the lateral geniculate nucleus of the cat (Bloomfield and Sherman, 1989). Shown are the relative voltages at various sites within the dendritic arbor and soma for a voltage change at a distal dendritic tip. For the relay cells, more than half the voltage change is propagated to the soma (Figure 3.6A,B,E), whereas for the interneurons, the propagation is less than 10% (Figure 3.6C–E). This difference between cell types is due to several factors: relay cells have larger diameter dendrites, they have less branching, and their branches tend more to match impedance among branch participants, leading to less voltage loss. As a result, synaptic locations within a relay cell's dendritic arbor, even at the most distal locations, can have a significant influence on the cell's firing properties, whereas more distally located synapses on interneurons are likely to have relatively little influence on firing (Cox et al., 1998).

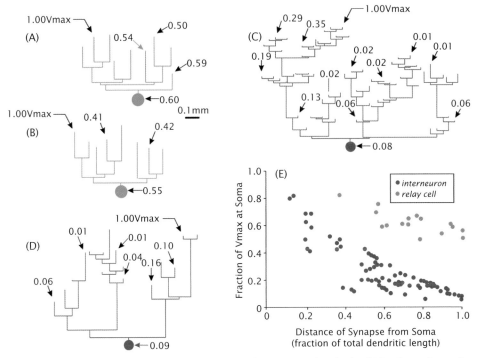

FIGURE 3.6 Cable modeling of the voltage attenuation that occurs within the dendritic arbors of two relay cells (blue) A,B and of two interneurons (red) C,D after the theoretical activation of a single synapse (i.e., of a single current injection leading to a relative depolarization as indicated). The cells from the cat's lateral geniculate nucleus were labeled by intracellular injection of a dye in vivo, and the stick figures represent a schematic view of one primary dendrite from each cell with all of its progeny branches. Each branch length is proportional to its calculated electrotonic length, and the shorter the length, the less current leaks out of the cell and thus the less voltage attenuates. The site of a theoretical voltage injection is indicated by the arrow at a dendritic tip labeled 1.00 Vmax (maximum voltage). Computed voltage attenuation at various dendritic endings within the arbor and soma is indicated by arrows and given as fractions of Vmax. E. Attenuation at the soma of a single theoretical voltage injection placed at different dendritic endings within the dendritic arbor as a function of anatomical distance of the voltage injection from the soma. Each voltage injection roughly mimics the activation of a single synapse. The abscissa represents relative anatomical distances normalized to the greatest extent of each arbor, and the plotted points represent values from the four cells shown in A–D. Redrawn from Bloomfield and Sherman, 1989.

3.3.2 Voltage-Dependent Channels

Thalamic neurons, including relay cells, interneurons, and thalamic reticular cells, have all the usual voltage-dependent conductances (McCormick and Huguenard, 1992), obviously including those underlying the action potential. What is relatively unusual for these cells is that they all have a sufficiently dense aggregate of dendritic T-type Ca^{2+} channels to activate low-threshold Ca^{2+} spikes and thereby exhibit burst firing (Jahnsen and Llinás, 1984; Huguenard and McCormick, 1994; Zhan et al., 1999; Sherman, 2001; Llinás and Steriade, 2006). The voltage and temporal properties of the T-channel differ slightly among relay cells, interneurons (Pape and McCormick, 1995) and thalamic reticular cells (Huguenard and Prince, 1992), but the basic behavior of burst versus tonic firing is the same. Figure 3.7 shows some of these properties for relay cells.

Figure 3.7A,B shows how a relay cell responds to a depolarizing input (e.g., an EPSP or current injection) in tonic versus burst firing mode (Zhan et al., 1999, 2000). When the resting membrane potential is sufficiently depolarized in amplitude and time (Figure 3.7A), the T-channels are inactivated (see Figure 3.4) and play no role. Here, a suprathreshold depolarizing input directly evokes single action potentials that continue as long as the depolarization remains above threshold. When the resting membrane potential is sufficiently hyperpolarized in amplitude and time (Figure 3.7B), the T-channels are de-inactivated. Now, the *same* depolarizing pulse evokes a large, low-threshold spike, which, in turn, evokes a burst of four action potentials. Two key differences exist between tonic and burst firing, among others. (1) Tonic firing persists as long as depolarization remains above threshold. In contrast, burst firing terminates, even with continued depolarization. This is because of inactivation and K^+-dependent repolarization of the Ca^{2+} channels that underlie the action potentials (see Figure 3.3B) and the continued depolarization of the injected current that prevents de-inactivation of the Ca^{2+} channels. (2) During tonic firing, the depolarizing input (e.g., an EPSP) *directly* evokes action potentials, but during bursting, the depolarizing input evokes action potentials *indirectly* via the low-threshold spike. Perhaps the most important point about these firing modes is that the exact same depolarizing input to the cell (whether a depolarizing current injection or an EPSP) will evoke a different firing pattern depending on the mode. In other words, the message relayed to cortex is different depending on the voltage history of the relay cell.

Figure 3.7C shows that, for a given level of de-inactivation of the T-channels due to a given sustained hyperpolarized membrane potential, the amplitude of the evoked low-threshold spike is invariant. This is because that spike is all-or-none, just like the conventional action potential, and so its evoked amplitude under such conditions is fixed. A fixed amplitude, low-threshold spike will evoke a fixed number of action potentials.

However, as indicated by Figure 3.7D, this does not mean that all bursts have a fixed amplitude and number of action potentials. As indicated by Figure 3.4B, the more the initial hyperpolarization, the more Ca^{2+} channels become de-inactivated, and more of these channels will evoke a larger low-threshold spike; thus the more the initial hyperpolarization, the larger the Ca^{2+} spike, and the more action potentials evoked.

Figure 3.7E shows another key difference between tonic and burst firing. During tonic firing (blue curves), the action potentials are directly activated by a suprathreshold depolarizing input such as an EPSP, and thus there is a monotonic relationship that is nearly linear for suprathreshold depolarization between the amplitude of the initial depolarization and the number of action potentials evoked. However, during burst firing (red curves), the relationship is markedly nonlinear, much like a step function, because a subthreshold depolarization evokes nothing, but, once suprathreshold, the Ca^{2+} spike evoked is relatively fixed in amplitude (Figure 3.7C), and thus a fixed number of action potentials is evoked, no matter how large the activating suprathreshold depolarization.

Burst and tonic firing modes are often referred to as completely different firing modes, but there is a sort of intermediate stage. As indicated in Figure 3.4B, if a relay cell is held at a level more depolarized than about −65 mV, all T-type Ca^{2+} channels are inactivated, and the neuron's response is strictly in tonic mode. If the neuron is held sufficiently hyperpolarized

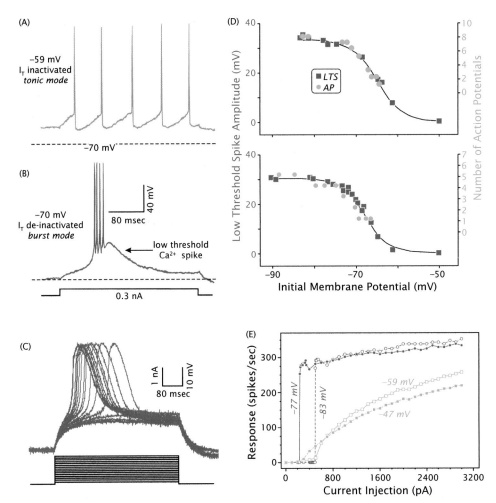

FIGURE 3.7 Properties of I_T and the low threshold Ca^{2+} spike. All examples are from relay cells of the cat's lateral geniculate nucleus recorded intracellularly in an in vitro slice preparation. A,B. Voltage dependency of the low Ca^{2+} threshold spike. Responses are shown to the same depolarizing current pulse delivered intracellularly but from two different initial holding potentials. When the cell is relatively depolarized (A), I_T is inactivated, and the cell responds with a stream of unitary action potentials as long as the stimulus is suprathreshold for firing. This is the *tonic mode* of firing. When the cell is relatively hyperpolarized (B), I_T is de-inactivated, and the current pulse activates a low threshold Ca^{2+} spike with four action potentials riding its crest. This is the *burst mode* of firing. C. All-or-none nature of low threshold Ca^{2+} spikes measured in the presence of tetrodotoxin (TTX), which blocks Na^+ channels, thereby preventing action potentials, in another geniculate neuron that was initially hyperpolarized, and current pulses were injected into the neuron starting at 200 pA and incremented in 10 pA steps. Smaller (subthreshold) pulses led to pure resistive-capacitative responses, but all larger (suprathreshold) pulses led to a low threshold spike. Much like conventional action potentials, the low threshold spikes are all the same amplitude regardless of how far the depolarizing pulse exceeded activation threshold, although there is latency variability for smaller suprathreshold pulses. D. Voltage dependency of amplitude of low threshold spike and burst response. Examples for two neurons are shown, and the upper is the same as shown in C. The more hyperpolarized the cell before being activated (*Initial Membrane Potential*), the larger the low threshold spike (*red squares* and *curve*) and the more action potentials (*AP*) in the burst (*blue circles*). The number of action potentials were measured first and then TTX was applied to isolate the low threshold spike for measurement. E: Input-output relationship for another cell. The input variable is the amplitude of the depolarizing current pulse, and the output is the firing frequency of the cell. To compare burst and tonic firing, the firing frequency was determined by the first six action potentials of the response, since this cell usually exhibited six action potentials per burst in this experiment. The initial holding potentials are shown, and –47 mV and –59 mV reflects tonic mode, whereas –77 mV and –83 mV reflects burst mode. Redrawn from Sherman and Guillery, 2001.

to bring an adequate number or density of these Ca^{2+} channels into play to activate a low-threshold spike, burst firing ensues. However, at a less hyperpolarized level, some of these Ca^{2+} channels will be de-inactivated, but their density is insufficient to activate an all-or-none low-threshold spike, so instead their activation by a depolarizing input such as an EPSP will evoke a relatively small nonpropagating depolarization that will affect the neuron's responsiveness by moving the membrane potential closer to the action potential threshold, thereby making the neuron more excitable (Deleuze et al., 2012; Alitto et al., 2019).

3.4 Intrinsic Properties of Cortical Neurons
3.4.1 Cable Properties
Cable modeling for cortical neurons has largely been limited to pyramidal cells in layers 2/3 and 5, and, even here, such modeling has been quite limited in relevance. This is because these cells have a rich array of active ionic channels in their dendrites, and activation of these channels usually trumps the relevance of the neuron's cable properties. What modeling has been done suggests that cortical pyramidal cells are quite compact electronically (Jack et al., 1975; Larkman et al., 1992). A dominant principle of cable modeling is that, the further a synapse is from the soma, the less effective will be its postsynaptic effect due to loss of current across the membrane as it flows to the soma. However, more recent evidence from pyramidal cells shows that active membrane properties can compensate for synaptic location to make synapses more or less equally effective regardless of their dendritic location (Hirsch et al., 1995; Magee and Cook, 2000; Branco and Hausser, 2011). Therefore, it seems that understanding how synaptic inputs integrate to affect cortical cells' firing depends much more on the neuron's active dendritic properties than their passive ones, at least for pyramidal cells.

3.4.2 Voltage-Dependent Channels
Numerous active ion channels have been discovered in cortical neurons, and these affect the firing properties of the neurons. Some have been identified based on responses to suprathreshold current injection. Cortical cells tend to respond at either very high frequencies ("fast spiking" cells) or lower frequencies ("regular spiking" cells). The former are thought to be GABAergic interneurons, whereas the latter are thought to be excitatory pyramidal cells. Fast spiking cells have very brief action potentials because their Na^+ channels repolarize quickly, and so these channels de-inactivate sooner, allowing for higher frequency firing. Pyramidal cells can be further divided into two response types. One fires action potentials at a fairly fixed frequency, whereas the other fires at a decreasing frequency, a pattern known as *spike-frequency adaptation*. The latter is due mainly to the buildup of an after-hyperpolarization due to opening of K^+ channels that increases as the response progresses. This further hyperpolarization causes the progressive reduction in the frequency of firing.

Some pyramidal cells in layers 2/3 and 5 exhibit voltage- and time-dependent Ca^{2+} conductances that lead to burst firing, qualitatively quite similar to that seen in thalamic relay cells, except that both low-threshold and high-threshold Ca^{2+} channels can be involved (Larkum et al., 1999, 2007; Llano and Sherman, 2009; Suzuki and Larkum, 2020). In layer 5, the largest pyramidal cells that project subcortically and send apical dendrites to layer 1 show this property, whereas the smaller layer 5 pyramidal cells that have no subcortical projections fail to exhibit this bursting property (Llano and Sherman, 2009). The higher threshold Ca^{2+} channels are often activated by back propagation of action potentials, and it has been demonstrated that, whereas synaptic inputs to layer 1 dendrites of these pyramidal cells alone have little effect on the cell's firing, the pairing of these layer 1 inputs with a back-propagating action potential can evoke a Ca^{2+} spike and bursting (Larkum et al., 2007).

3.5 Summary and Concluding Remarks

Ultimately, understanding the functioning of the neuron is a sine qua non for understanding functioning of the brain, writ large. We are far from that goal, and perhaps the best we can claim is a beginning of an understanding of the underlying complexities involved. Since the code of neural communication is the pattern of action potentials generated by each neuron, we need to understand how intrinsic properties of neurons shape these patterns. Generally, these patterns are the product of transmembrane voltage changes at the axon hillock. In turn, these changes are the product of quite complex interactions of a neuron's passive cable properties plus the distribution and properties of many voltage- and time-dependent ion channels found throughout a neuron's membrane. To further complicate matters, these passive and active membrane properties are highly variable, not only among neurons, but also within neurons as plastic or homeostatic actions alter them. Clearly, one of the great challenges of neuroscience is a better holistic understanding of neuronal single-cell properties.

3.6 Some Outstanding Questions

1. How do voltage- and time-dependent conductances interact with one another and with the cell's cable properties to affect how the cell responds to various inputs?
2. What is the complete pattern in dendrites of different thalamic and cortical neurons of voltage- or other dependent channels?
3. What effect do these dependent channels have on synaptic integration?
4. Under what conditions and for which cells does back-propagation of action potentials occur, and, if so, what is the significance of this for synaptic integration?
5. Given the apparent importance of voltage-dependent conductances seen with in vitro methods, what is the range of membrane potentials typically seen under physiological conditions in thalamic and cortical neurons cells of awake, behaving animals?
6. Are there important differences in different classes of thalamic neuron regarding their intrinsic properties, especially their voltage- and time-dependent ionic channels?

4

4

Synaptic Properties

Synapses are the means by which neurons communicate with one another. The vast majority of synapses work via chemical transmission of signals, but some connect neurons via direct electrical means. These electrical synapses are considered briefly below. In human cortex, there are roughly 2×10^{10} neurons, each of which is innervated by 1 to 10×10^3 synapses, so there are approximately $2 \times 10^{13-14}$ synapses in cortex alone. Clearly, this leads to quite complex circuitry with a mind-boggling number of possible combinatorial configurations. Indeed, because of these numbers, some have called the brain the most complex machine in the universe, although our overall ignorance of the universe makes such an assertion seem an exercise in hubris.

What sets the brain apart from most other organs is this functionally critical connectivity. If one understands the working of a cardiac or renal cell, one understands much of the functioning of the heart or kidneys. Understanding the functioning of a neuron is a nice starting point, but this alone provides precious little understanding of the brain. What is lacking is the understanding of the pattern of synaptic connections that underlies the real power of the brain. Thus, the synapse represents a particularly key element that actually sets the brain apart from other organs. In what follows, we provide a brief introduction to the functioning of the synapse, particularly as it applies to thalamus and cortex.

4.1 Chemical Synapses

An electron micrograph of a typical chemical synapse is shown in Figure 4.1 and its basic organization in Figure 4.2. A presynaptic neuron (gray in Figure 4.2A) fires an action potential that travels down the axon to terminals that form synapses onto a postsynaptic neuron (blue). A blow up of one such synaptic terminal is shown in Figure 4.2B. The action potential depolarizes the terminal which, in turn, activates Ca^{2+} channels, leading to the flow of Ca^{2+} into the terminal. In a series of events beyond the scope of this account (but for more detail, see Pickel and Segal, 2013; Dittman and Ryan, 2019), the increased Ca^{2+} leads to the synaptic vesicles fusing with the membrane and dumping their contents into the synaptic cleft. These

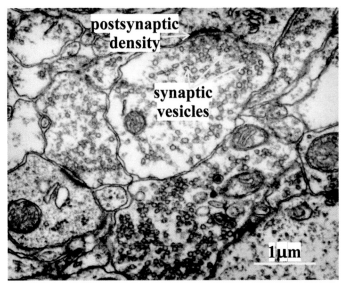

FIGURE 4.1 Synapse in cortex of a mouse. The presynaptic terminal is filled with synaptic vesicles, particularly near the site of the synapse, which is marked by postsynaptic density on the target dendrite. This density is likely an image of postsynaptic receptors. Image kindly provided by N. Kasthuri.

contents are the neurotransmitter molecules, and these molecules bind with postsynaptic receptors[1] to evoke opening or closing of various postsynaptic ion channels; this, in turn, produces a postsynaptic potential, an excitatory postsynaptic potential (EPSP) or an inhibitory one (IPSP).

There are many neuroactive substances that can be released from synaptic terminals and can thus be regarded as neurotransmitters, but the major ones discussed here are glutamate, GABA, acetylcholine, noradrenaline (also known as norepinephrine), serotonin, and dopamine. Neurons using one of these neurotransmitters are often described in this context: thus we have glutamatergic, GABAergic, cholinergic, etc., neurons.

4.1.1 Glutamate

Glutamate is the major neurotransmitter used to transfer information either between neurons locally or between brain regions more globally. It is normally thought to act as an excitatory neurotransmitter, because its binding to postsynaptic receptors usually opens sodium (Na^+) and often Ca^{2+} channels, leading to an EPSP, but, as described below in our discussion of metabotropic postsynaptic receptors, it can also act to inhibit its postsynaptic target. Glutamatergic neurons typically project in a highly topographic manner. Some glutamatergic neurons project over relatively long axons to other brain regions and others

---

1. The term "receptors" can sometimes confuse, because it has two meanings in neuroscience. One refers to specialized cells that act to transduce environmental energy into a neural code. Examples are rods and cones in the retina, Pacinian corpuscles in the skin and joints, and hair cells in the inner ear. The other meaning refers to specialized proteins in the postsynaptic membranes that react with neurotransmitters to produce postsynaptic effects like EPSPs or IPSPs. "Receptors" as used here refer exclusively to these postsynaptic proteins.

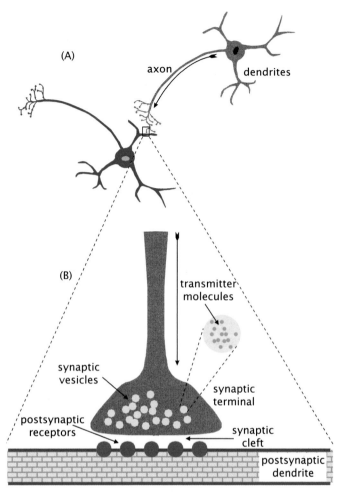

FIGURE 4.2 Schematic view of a synapse (refer also to Figure 4.1). A. A presynaptic neuron (gray) innervates another neuron (blue), and an action potential initiated in the former will travel down the axon to synaptic terminals innervating the latter. B. Blowup of region shown in A. Depolarization of the terminal from the action potential will lead to an increase of internal Ca^{2+} that will cause the synaptic vesicles (yellow circles) to fuse with the membrane and discharge their contents (i.e., neurotransmitter molecules) into the synaptic cleft to interact with postsynaptic receptors (red circles).

act as local interneurons, but even the long-projecting neurons contribute to local circuitry with axonal branches.

Glutamatergic neurons are especially important in the functioning of the brain generally and thalamus and cortex specifically. Chapter 5 expands on the subject of glutamatergic synapses by describing how they can be classified into two very different types, "driver" and "modulator." Here, we limit discussion to a more general description of synaptic function.

4.1.2 GABA

GABAergic neurons in thalamus and cortex are almost exclusively local, inhibitory neurons. We consider neurons of the thalamic reticular nucleus among these, because, whereas they do project out of the reticular nucleus to various thalamic relay nuclei, the projections are still limited to thalamus. GABAergic neurons act via postsynaptic receptors by opening Cl⁻ or K⁺ channels, typically leading to an IPSP.[2]

4.1.3 Classical Modulatory Neurotransmitters

Classical modulatory neurotransmitters for thalamus and cortex include acetylcholine, noradrenaline, serotonin, and dopamine. As noted above, information being processed by thalamus and cortex is thought to be carried between neurons and areas by glutamatergic neurons, although, as noted and described more fully in Chapter 5, a class of glutamatergic synapses functions like classical modulatory synapses rather than serving as routes of information transfer. Modulatory circuits act not so much to carry information for processing but rather to modulate how that information is processed. For example, modulatory inputs to thalamus or cortex can affect the strength of glutamatergic synapses, affect the balance between excitation and inhibition in circuitry there, and underlie synaptic plasticity, among other actions.

Neurons using these neurotransmitters reside in neither thalamus nor cortex, which means that, like many glutamatergic neurons, they operate via long-range axons. Figure 4.3 summarizes the locations of neurons of origin of these systems as well as their projection patterns. As shown, these modulatory projections are generally quite diffuse and non-topographic. Thus, the modulatory effects they produce tend to be related to overall behavioral states like drowsiness, alertness, etc. These projection patterns contrast sharply with those of glutamatergic projections, which tend to be quite topographic, such as the highly retinotopic retinogeniculate or geniculocortical projections in the visual system.

Cholinergic modulation is particularly important for thalamic and cortical functioning. As indicated in Figure 4.3A, the distribution of cholinergic pathways is quite widespread in the brain (Li et al., 2018). Cholinergic input to thalamus emanates from the brainstem in scattered, loosely organized groups under a variety of confusing names, such as inclusion in the "brainstem reticular formation," the "pedunculopontine nuclei," and the "parabrachial region" (for review, see Jones, 2007; Sherman and Guillery, 2013). Cholinergic input to cortex derives from other scattered neuron groups commonly referred to as the "basal forebrain."

4.1.4 Other Neurotransmitters

There are several other neuroactive substances that generally are co-localized in terminals with the more common neurotransmitters noted above. A major type is a variety of neuropeptides, which are short chains of amino acids (Van den Pol, 2012; Devi and Fricker, 2013).

---

2. There is some terminological confusion about the action of certain GABAergic neurons as possibly acting in an excitatory manner. Normally, the reversal potential for Cl⁻ is about −80 mV, and so for a neuron resting at, say, −70 mV, GABA responses will cause Cl⁻ to flow into the neuron, producing an IPSP. However, in some cases (e.g., often during development) the internal concentration of Cl⁻ can be sufficiently high that its reversal potential can be slightly depolarized with respect to the neuron's resting potential (e.g., at −60 mV reversal potential for a neuron resting at −70 mV). Now the activity of GABA would produce a small EPSP as Cl⁻ flows out of the neuron. Because of an EPSP, some call this an excitatory effect. However, to evoke an action potential requires depolarization to about −55 mV, and, with these Cl⁻ channels open, any attempt to depolarize beyond −60 mV would be resisted as the action here would be to clamp the membrane potential to −60 mV. Acting to prevent production of an action potential, even with an initial EPSP, should be seen as an inhibitory action.

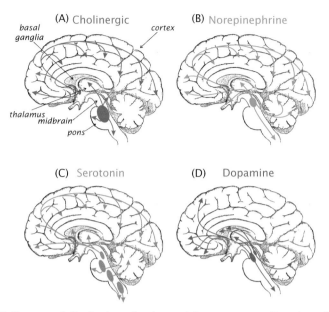

FIGURE 4.3 A–D: Source and distribution of major modulatory systems. Note the rather diffuse and non-topographical nature of the innervation in thalamus and cortex.

These are found quite commonly among cortical GABAergic interneurons; the peptide type is often used to classify these neurons. These peptides are released along with the classical neurotransmitter with which they co-exist. However, the firing pattern typically dictates the relative amount of each that is released: for example, often peptides require higher firing rates for release.

Finally, the gas nitric oxide often coexists with acetylcholine and has neuroactive effects as a diffusible gas that can enter cells and cause numerous modulatory changes, including modulation of ion channel activity (Zhang and Snyder, 1995; Dawson and Dawson, 1996; Hardingham and Fox, 2006; Garthwaite, 2016).[3] By diffusing into cells, it does not operate by activating postsynaptic receptors. Also, by affecting smooth muscle fibers, it can increase blood flow locally in a brain area and thereby affect neuronal activity (Umans and Levi, 1995).

4.2 Electrical Synapses

In both cortex and thalamus, some GABAergic neurons, in addition to receiving synaptic inputs using chemical transmission, also form gap junctions between one another to create another means of neuron to neuron communication. These gap junctions are also known as *electrical synapses*, because they allow current to flow unabated between neurons and play a role in synchronizing neuronal firing. Gap junctions have been found in neurons of the thalamic reticular nucleus (Liu and Jones, 2003; Landisman and Connors, 2005; Lam et al.,

---

3. Nitric oxide was declared "molecule of the year" by *Science* in 1992 (Koshland, Jr., 1992), perhaps because nitric oxide is a key component of Viagra and related drugs.

2006) and certain GABAergic cortical neurons (Kandler and Katz, 1998; Liu and Jones, 2003; Caputi et al., 2009; Ma et al., 2011). The efficacy of some gap junctions can be modulated by voltage changes or the action of modulatory inputs (Landisman and Connors, 2005; Haas and Landisman, 2011; Papasavvas et al., 2020).

4.3 Postsynaptic Receptors

Chemical synapses work by releasing neurotransmitter from a presynaptic terminal, allowing the neurotransmitter molecules to diffuse across the narrow synaptic cleft to bind to postsynaptic receptors, which, in turn, leads to postsynaptic events like EPSPs or IPSPs. Each neurotransmitter can react with a variety of different postsynaptic receptors, and the particular receptor type associated with a given synapse plays a major role in the action of the synapse. For example, we normally think of glutamate and acetylcholine as excitatory neurotransmitters, but the nature of the postsynaptic receptor can turn each into an inhibitory neurotransmitter. Thus, we no longer think of a neurotransmitter as excitatory or inhibitory but rather see that classification as more related to the receptor type.

4.3.1 Ionotropic and Metabotropic Receptors

Postsynaptic receptors can be divided into two main types: ionotropic and metabotropic. Most neurotransmitters are associated with both types, often at the same synapse. Figure 4.4 illustrates the main differences between these receptors.

4.3.1.1 Ionotropic Receptors

Ionotropic receptors are simpler in functional organization (Figure 4.4A), and details of their function can be found elsewhere (Catterall, 2010). The receptor molecule itself includes the relevant ion channel and is a complex monomeric transmembrane protein comprised of several subunits that may combine in different ways to subtly affect receptor functioning. The receptor protein molecule wraps back and forth across the membrane several times. The channel is normally blocked, but binding to the neurotransmitter causes a conformational change in the receptor that exposes the ion channel, allowing movement of the relevant ion across the neuronal membrane. For glutamate (Figure 4.4Ai), the relevant ionotropic receptors are α-amino-3-hydroxy-5-methyl-4-isoxazole propionic acid (AMPA), N-methyl-D-aspartic acid (NMDA), and sometimes kainate. When glutamate binds to one of these receptors, Na^+ and often Ca^{2+} channels open, allowing these ions to flow into the neuron, creating an EPSP. For GABA (Figure 4.4Aii), the ionotropic receptor is $GABA_A$. Here, the conformational change leads to opening of a Cl^- channel, and flow of Cl^- into the neuron creates an IPSP. Partly because of the direct linkage of the receptor to the ion channel, the response is fast, with a latency of 1 msec or less and a duration of 10 msec or so.[4]

---

4. The underlying synaptic conductance changes lead to current flow across the membrane. Because of the resistive-capacitive properties of the membrane, the voltage changes seen across the membrane last longer

(A) Ionotropic Receptor (e.g., for AMPA or GABA$_A$)

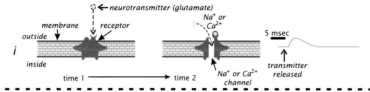

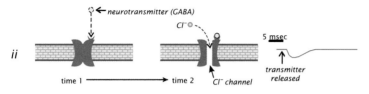

(B) Metabotropic Receptor

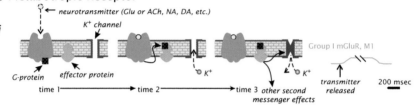

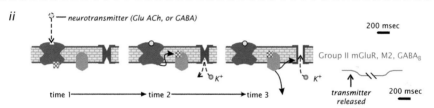

FIGURE 4.4 Schematic depiction of ionotropic and metabotropic receptors, each type shown repeatedly at different times (*time 1* and *time 2* for the ionotropic receptors, and *time 1, time 2,* and *time 3* for the metabotropic receptors). A. For the ionotropic examples, *time 1* represents the period before binding to the neurotransmitter, and *time 2* is the period after binding. The binding causes a conformational change that opens the ion channel, which forms the central core of the receptor complex. B. For the metabotropic receptors, *time 1* is the period before neurotransmitter binding, and just after binding (*time 2*) a G-protein is released, which reacts with an effector protein to produce a cascade of biochemical reactions eventually resulting in opening or closing of an ion channel, usually a K$^+$ channel (*time 3*). Further details in the text.

Abbreviations: *ACh*, acetylcholine; *AMPA*, α-Amino-3-hydroxy-5-methyl-4-isoxazole propionic acid; *DA*, dopamine; *GABA*, γ-aminobutyric acid; *Glu*, glutamate; *M1 & M2*, muscarinic receptors; *mGluR*, metabotropic glutamate receptor; *NA*, noradrenalin. (Redrawn from Sherman and Guillery, 2013).

than the current flow, and this relates to the membrane time constant. Thus, measures of synaptic events recorded with the technique of *voltage clamp*, which measures postsynaptic currents (e.g., EPSCs), produce briefer events than the same events measured during *current clamp*, which measure postsynaptic voltages (e.g., EPSPs).

Although not shown in Figure 4.4A, nicotinic receptors are ionotropic receptors for acetylcholine, and activation of these leads to EPSPs. These receptors play an important role in thalamic and cortical functioning.

4.3.1.2 Metabotropic Receptors

The postsynaptic events involving metabotropic receptors are more complex (Nicoll et al., 1990; Pin and Bockaert, 1995; Pin and Duvoisin, 1995; Conn and Pin, 1997). Figure 4.4B demonstrates this and should be compared to the events depicted in Figure 4.4A for ionotropic receptors. Metabotropic receptors themselves are simple proteins that usually are made up of a single polypeptide, but they represent the start of a complicated chain of events that eventually evoke postsynaptic responses. Once the neurotransmitter binds to a metabotropic receptor, a series of actions is triggered (see Figure 4.4B). First is a conformational change in the receptor that leads to the release of a G-protein, which transposes to react with a nearby effector protein, and this leads to a cascade of biochemical reactions in the postsynaptic neuron, a process known as a *second-messenger pathway*. These reactions in turn lead to several changes in the neuron. One such change is opening or closing of specific ion channels. In the case of metabotropic receptor activation in the thalamus and cortex, the main effect is on K^+ channels, although Ca^{2+} and other channels can also be affected (Gereau and Conn, 1994; Anwyl, 2009; Tsanov and Manahan-Vaughan, 2009; Luscher and Huber, 2010). Note that effects on K^+ and other channels are but one of the results of the second messenger cascade of chemical reactions. That is, some results can affect a range of metabolic activities and even gene expression, although these and other possible effects on thalamic and cortical neurons remain largely unexplored.

A set of metabotropic receptors leads to EPSPs by closing K^+ channels (Figure 4.4Bi). Other metabotropic receptors operate by opening K^+ channels (Figure 4.4Bii), thereby producing IPSPs. The signaling pathways for the different types of response indicated in Figure 4.4Bi and Bii generally involve different G-proteins (Gq for Figure 4.4Bi and Gi/o for Figure 4.4Bii) and other features of the second-messenger cascade (for further details, see Coutinho and Knopfel, 2002; Kim et al., 2008).

As indicated in Figure 4.4, postsynaptic potentials produced in this way have a much longer time course than that seen with ionotropic receptors, with a latency of 10 msec or more and a duration of hundreds of msec to several sec or more (Govindaiah and Cox, 2004). The longer latency and duration of postsynaptic potentials is presumably due to the temporal requirements and duration of the second-messenger events. This difference in duration of postsynaptic potentials seems to play a role in control of voltage- and time-gated ion channels (reviewed in Sherman and Guillery, 2013). A good example of this is the T-type Ca^{2+} channel noted in Chapter 3. Recall that inactivation of this channel requires sufficient depolarization that must be sustained for at least 100 msec or so, and, likewise, de-inactivation requires sufficient hyperpolarization that must be sustained for at least 100 msec or so (Figure 3.3B). Activation of ionotropic receptors evokes postsynaptic potentials that are typically too brief to affect the inactivation state of these channels, but postsynaptic potentials produced by activation of metabotropic receptors fit the bill nicely. Indeed, evidence exists that activation of metabotropic glutamate receptors serves to affect the inactivation state of these T-type Ca^{2+}

channels (Godwin et al., 1996). As stated in Chapter 3, these inactivation time constants for the T-type Ca^{2+} channel apply to most other voltage-gated channels, and so activation of metabotropic receptors would seem ideally suited in general for controlling voltage- and time-gated ion channels.

4.3.1.3 Metabotropic Glutamate Receptors

Particularly important to thalamic and cortical functioning are metabotropic glutamate receptors. There are eight different types of these receptors recognized in the brain, and these are distributed into three groups. Activation of Group I (types 1 and 5) metabotropic glutamate receptors leads to prolonged EPSPs (Figure 4.4Bi), whereas activation of Group II (types 2 and 3) receptors leads to prolonged IPSPs (Figure 4.4Bii). Group III metabotropic glutamate receptors (types 4 and 6–8) have not been much studied in thalamus or cortex and are not further considered (but see Salt and Eaton, 1995; Gu et al., 2012).

4.3.1.4 Muscarinic Receptors

Cholinergic inputs can also activate metabotropic receptors known as muscarinic receptors. Like metabotropic glutamate receptors, these come in various types, and the important ones for thalamic and cortical functioning are known as M1 and M2 receptors. M1 receptors are associated with closure of K^+ channels, leading to an EPSP (Figure 4.4Bi), whereas M2 receptors are associated with opening of K^+ channels, leading to an IPSP (Figure 4.4Bii).

4.3.1.5 $GABA_B$ Receptors

$GABA_B$ receptors are the metabotropic receptors for GABA and produce IPSPs by opening K^+ channels (Figure 4.4Bii). There are two important differences between activation of $GABA_A$ and $GABA_B$ receptors. One already noted is the time course: activation of $GABA_A$ receptors leads to a faster, briefer IPSP. The other has to do with the ion involved: because Cl^- has a less hyperpolarized reversal potential than does K^+, $GABA_B$ receptor activation tends to produce a larger IPSP.

4.3.1.6 Other Metabotropic Receptors

Noradrenaline, serotonin, and dopamine all operate through metabotropic receptors specific to each, although ionotropic receptors may also be involved.

4.3.1.7 Differences in Activation Requirements for Ionotropic and Metabotropic Receptors

As noted in Figure 4.5, whereas ionotropic receptors (red circles) are found within the synaptic zone, metabotropic glutamate receptors (green stars) tend to be located perisynaptically (Nusser et al., 1994; Lujan et al., 1996; Kennedy, 2000). This may explain why metabotropic glutamate receptors generally require higher levels of presynaptic activity for their activation than do ionotropic glutamate receptors, which can be activated by a single action potential; this difference presumably exists because more glutamate must be released presynaptically to reach the more distant metabotropic receptors. Nonetheless, activation requirements for

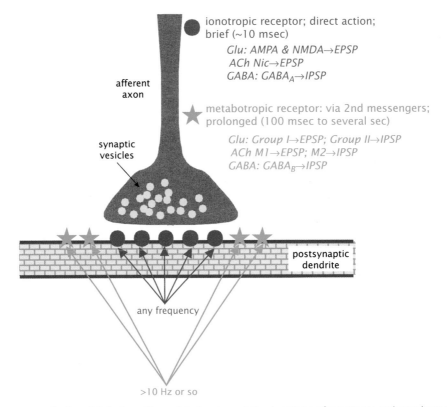

FIGURE 4.5 Further details from Figure 4.2 showing relative locations of ionotropic and metabotropic postsynaptic receptors (red circles and green stars, respectively).

Abbreviations: As in Figure 4.4 plus *EPSP*, excitatory postsynaptic potential; *IPSP*, inhibitory postsynaptic potential; *Nic*, nicotinic receptor.

metabotropic receptors are clearly within the normal physiological range: in a study of the activation of metabotropic glutamate receptors, it was found that they could be activated by firing rates as low as 10 spikes/sec, with higher firing rates achieving more activation (Viaene et al., 2013).

4.3.2 Paired-Pulse Synaptic Effects

There is a temporal factor to synaptic functioning because most synapses operate in a frequency-dependent manner (Thomson and Deuchars, 1994, 1997; Lisman, 1997; Chung et al., 2002). That is, the frequency or, more generally, the temporal pattern of action potentials arriving at the presynaptic terminal, affects the amplitude of evoked EPSPs or IPSPs. It should be noted that this feature is described in relation to ionotropic receptors only, and we shall describe it specifically for ionotropic glutamate receptors, but the general features apply as well to other ionotropic receptors, whether producing EPSPs or IPSPs.

Paired-pulse effects are calculated by comparing the size of a postsynaptic potential evoked by a first action potential to one evoked by the next action potential as a function of the intervening time interval. (One can also use later postsynaptic potentials in a longer train by comparing the sizes of the *n*th postsynaptic potential to that of the *n*th + 1.) This represents responses evoked by pairs of stimuli, and so this property is often called a *paired-pulse effect*. Figure 4.6 illustrates the two types of paired-pulse effects commonly observed. *Paired-pulse depression* (Figure 4.6A) occurs when the second postsynaptic potential is smaller than the first for interspike intervals briefer than a certain value. In contrast, *paired-pulse facilitation* (Figure 4.6B) occurs when the second evoked postsynaptic potential is larger. The effective interspike intervals for both depression and facilitation are similar, with time constants of several tens of milliseconds that vary among synapses; with longer intervals, there is no facilitation or depression (Figure 4.6C,D). Shown in Figure 4.6 are fairly typical examples,

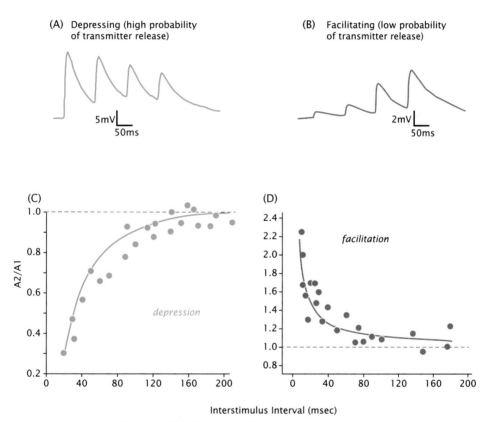

FIGURE 4.6 Examples of paired-pulse effects shown for glutamatergic synapses. The recordings are from single relay cells in slices of the mouse lateral geniculate nucleus in response to electrical stimulation of the neurons. A. Paired-pulse depression in response to retinal stimulation at 10 Hz. Note that during the train, EPSPs become smaller. Such a response is associated with a high probability of neurotransmitter release for each action potential invading a terminal. B. Paired-pulse facilitation in response to layer 6 corticogeniculate stimulation at 10 Hz. Note that during the train, EPSPs become larger. Such a response is associated with a low probability of neurotransmitter release. C. Time course of the paired-pulse depression in *A*, showing the ratio of amplitudes of the second EPSP divided by the first (A2/A1) as a function of interstimulus interval. D. Time course of paired-pulse facilitation in B, showing the ratio of amplitudes of the second EPSP divided by the first (A2/A1) as a function of interstimulus interval. Redrawn from Sherman and Guillery, 2013.

but there is considerable variability in the temporal properties and extent of these effects among neurons.

The explanation for these different paired-pulse effects remains somewhat unsettled; the consensus view is presented here. In this view, the mechanisms are largely presynaptic and relate to the fact that neurotransmitter release is a stochastic process. That is, for a given presynaptic terminal, an invading action potential will cause neurotransmitter release with a probability between zero and one (Dobrunz and Stevens, 1997), and this probability for any synapse can change depending on various factors such as changing intracellular Ca^{2+} concentrations and available neurotransmitter stores. Typically, the larger the Ca^{2+} entry evoked by the invading action potential, the greater the probability of release: neurotransmitter release increases with internal Ca^{2+} concentration nonlinearly as a power function with a power of 3 to 4 (Landò and Zucker, 1994). Note that most axons contribute multiple synapses to any postsynaptic target neuron, sometimes hundreds, and the release probability among these synapses from a single axon can vary greatly (Flight, 2008; Kavalali, 2015). Thus, for a given input from an axon to a postsynaptic neuron, the amount of neurotransmitter released due to an action potential will be related to the average of probabilities for each of its synapses. Thus, if the probability for release for each synapse from a given axon is, on average, 0.5, and the axon contacts a given neuron with, say, 50 synaptic contacts, then an action potential will cause an average of 25 synapses on that target to release neurotransmitter and produce a postsynaptic potential. The point is that, for any one pair of synaptically linked neurons, the larger the probability of release, the larger the resultant postsynaptic potential.

Given this, paired-pulse depression can be explained as follows. If synapses from an axon have a high average probability of release (e.g., $p > 0.5$), an action potential will result in much neurotransmitter being released, leading to a significant reduction in neurotransmitter stores in the terminals. If another action potential arrives before these stores can be replenished, less neurotransmitter is released by this following action potential: thus the first EPSP is larger than the second, and we have paired-pulse depression. As Figure 4.6C shows, if enough time elapses between the first and second action potential, the neurotransmitter stores are replenished and no depression is seen.

A different explanation is offered for paired-pulse facilitation. This occurs when the probability of release is low (e.g., $p < 0.5$), which results from a relatively small Ca^{2+} conductance evoked by the action potential. When a single action potential and the subsequent Ca^{2+} influx fail to promote neurotransmitter release, the Ca^{2+} concentration will remain elevated for some time. If a second action potential follows the first while the internal Ca^{2+} concentration remains elevated, it will cause a second wave of Ca^{2+} entry that will sum with what remains from the first, much like temporal summation in postsynaptic potentials. As noted above, neurotransmitter release increases rapidly with internal Ca^{2+} concentration, and so the result will be a higher probability of neurotransmitter release for the second action potential, leading to a second EPSP that is larger than the first, or paired-pulse facilitation. Again, the time interval between action potentials is key: if the second action potential does not follow soon enough, the Ca^{2+} concentration raised by the first action potential will have decayed to resting levels (Figure 4.6D).

Note that this explanation for paired-pulse effects is based on presynaptic factors involving probability of release and size of neurotransmitter stores. However, postsynaptic factors such as receptor desensitization may also play a role, perhaps even a dominant one. Whatever the explanation for paired-pulse facilitation or depression, the fact that effectively all synapses onto thalamic and cortical neurons either depress or facilitate underscores the importance of these phenomena in circuit functioning. For a synapse showing paired-pulse facilitation, afferent firing rates too low to evoke this facilitation would be relatively ineffective in influencing the postsynaptic neuron.

One interesting suggestion for paired-pulse depression is that it provides a gain control mechanism for synaptic processing (Chung et al., 2002; Priebe and Ferster, 2012). That is, the depression reduces EPSP amplitudes monotonically with firing rates. The lower amplitudes at higher rates help to avoid saturation of the postsynaptic response, thereby extending the dynamic range of afferent frequency over which the synapse functions.

4.4 Presynaptic Receptors

It has long been appreciated that many presynaptic terminals throughout the central nervous system have receptors for various transmitters (Thompson et al., 1993; Wu and Saggau, 1997; Miller, 1998; MacDermott et al., 1999; Kullmann, 2001; Vitten and Isaacson, 2001). This includes many terminals in thalamus and cortex, and the receptors identified respond to virtually all of the neurotransmitters noted above, including glutamate (Porter and Nieves, 2004; Brasier and Feldman, 2008; Govindaiah et al., 2010; Beierlein, 2014; Groleau et al., 2015; Banerjee et al., 2016; Barre et al., 2016; Park et al., 2017; Sottile et al., 2017). It is thought that activation of these receptors influences transmitter release from the terminal, typically by affecting the magnitude of Ca^{2+} that enters the terminal following an invading action potential. Interestingly, virtually all of these receptors are metabotropic, again including metabotropic glutamate receptors. Two exceptions are presynaptic nicotinic receptors (Disney et al., 2007; Sottile et al., 2017) and NMDA receptors (Banerjee et al., 2016). Typically, the metabotropic effects are to open K^+ channels, which reduces the Ca^{2+} influx associated with an action potential and thereby reduces neurotransmitter release (DePasquale and Sherman, 2012, 2013; Govindaiah et al., 2012; Lam and Sherman, 2013).

With the exception of thalamic triadic arrangements described in Chapter 7, these receptors on terminals are not themselves postsynaptic to synaptic terminals, which means that the appropriate neurotransmitter must travel to these receptors over a relatively large distance; this is known as *volume transmission* (Fuxe et al., 2007). However, rather than being a passive movement of the neurotransmitter, there is evidence that specialized glial cells, *neurogliaform cells*, provide for this movement (Overstreet-Wadiche and McBain, 2015).

4.5 Summary and Concluding Remarks

Synapses are a main feature that sets the brain apart from other organs. Like neurons, cells in other organs often have complex properties such as regulated ion channels, but it is the synaptic connections that underlie the great complexity and functionality of the central nervous system.

It used to be thought that synaptic function could be largely understood on the basis of the neurotransmitter systems they employed, so that synapses could simply be defined as glutamatergic, cholinergic, GABAergic, etc. However, we now appreciate that there is much more complexity to the tale that underlies much of the brain's computational power and flexibility. In particular, postsynaptic receptors are now recognized as a much richer group allowing for a far wider range of synaptic functioning. This includes both ionotropic and metabotropic classes available to all the major neurotransmitter systems, and each of these often has subclasses that offer further extension of synaptic functionality. One result of different receptor types is that many neurotransmitters once thought of as purely excitatory, such as glutamate and acetylcholine, can provide postsynaptic inhibition based on the receptor types activated.

Metabotropic receptors are particularly interesting, because, via second-messenger systems, their activation not only affects ion channels leading to postsynaptic potentials but can also have many other effects on the postsynaptic cell, including effects on gene expression and long-term plasticity. This class of receptors has been recognized only relatively recently, and there is still much to learn about them. In particular, metabotropic glutamate receptors, which are widely distributed in thalamus and cortex, are poorly understood, especially in regards to their effects beyond opening or closing ion channels. This is one area that is especially in need of further research.

4.6 Some Outstanding Questions

1. Given that second-messenger processes are evoked by activation of metabotropic receptors, what are the long-term effects of such activation in thalamus or cortex beyond just producing ion channel opening or closing?

2. What is the functional significance of the mix of receptor types, especially ionotropic and metabotropic, related to many inputs to thalamic and cortical neurons? Do certain patterns of activity activate different receptors relatively selectively? Do individual axons in these circuits activate different combinations of receptors associated with the entire pathway? For instance, do all corticogeniculate axons from layer 6 activate both metabotropic and ionotropic receptors, or do some activate just one or the other?

3. How common is back propagation of the action potential that invades dendrites of thalamic or cortical neurons, and, to the extent that it occurs, how does it affect synaptic integration on these neurons?

4. Many circuits that convey signals related to behavioral state include presynaptic receptors. How do effects from presynaptic receptor activation interact with mechanisms involved with rate-dependent synaptic transmission?

5

Glutamatergic Drivers and Modulators

One of the first steps necessary to analyze a complex system is to provide a detailed classification of its component parts. Thus, early in the study of retina or cortex was the identification of the various types of neurons in each. For retina, we have photoreceptors, bipolar cells, ganglion cells, etc., and for cortex, pyramidal and stellate cells. But in both regions, each of these classes has important subclasses: for retina, rods and cones for photoreceptors and various ganglion cell classes; for cortex, both pyramidal and stellate cells have been classified into numerous subtypes.

Likewise, parsing the complex circuitry of the brain has been aided immensely by classifying the various circuit elements. A first pass has involved identifying afferent pathways by their neurotransmitters. This has led to the early notion that each neurotransmitter class subserves a specific function: glutamatergic inputs carry information to be acted upon; GABAergic inputs provide necessary inhibitory control of circuits to prevent purely excitatory connections from causing the circuit to "explode" due to positive feedback loops; and the various modulatory circuits modulate how glutamatergic inputs are processed in keeping with behavioral factors such as overall behavioral state.

However, it now seems clear that, just as retinal and cortical cell classes can be productively broken down into subclasses, so can the classification of circuitry based on neurotransmitters. This is especially true for glutamatergic inputs in thalamus and cortex, inputs that have been further divided into classes known as "drivers" and "modulators"[1] (Sherman and Guillery, 1998, 2013). The initial basis for this classification arose from an early appreciation of glutamatergic inputs to the cat's lateral geniculate nucleus.

---

1.   Recent publications have referred to the driver input as Class 1 and the modulator as Class 2 to avoid terminology suggestive of function (reviewed in Sherman and Guillery, 2013), which is not entirely established. For simplicity, in deference to terminology that now seems common and to reflect our current thinking, we use the "driver" and "modulator" terminology here.

5.1 Glutamatergic Inputs to Geniculate Relay Cells

Relay cells in the lateral geniculate nucleus, which provide the major relay of retinal input to cortex, receive two major glutamatergic inputs: one from the retina, and the other from layer 6 of V1, or primary visual cortex (Figure 5.1). One way to judge the possible information conveyed by a visual neuron is to consider how it responds to visual stimuli (i.e., its receptive field properties), because these properties represent the essential information about the visual world that such a neuron can convey (Hubel and Wiesel, 1959, 1961; Gilbert, 1977; Grieve and Sillito, 1995; Usrey et al., 1999). Retinal inputs are monocularly driven (i.e., appropriate visual stimulation of the eye of origin of a retinal axon will evoke responses in the axon, but stimulation of the other eye will not) and have a center-surround organization. The cortical input is quite different: the receptive fields of these cells tend to be binocularly driven and have specificity for visual stimuli, such as orientation and direction selectivity, not seen in retinal inputs. Because the geniculate relay cell has receptive field features that closely match those of its retinal input and look nothing like those of its cortical inputs, it is clear that it is the retinal input that provides the main information to be relayed to cortex.

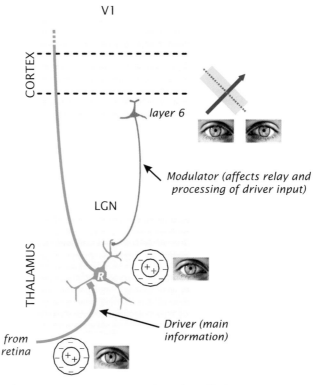

FIGURE 5.1 Main glutamatergic inputs to a geniculate relay cell. The retinal input has a monocular receptive field with a center-surround organization, whereas the cortical input has a binocular receptive field with orientation and often direction selectivity. The geniculate relay cell's receptive field closely matches that of its retinal input and looks nothing like its cortical input. See text for details.

If the retinal input serves to bring information to geniculate relay cells, what is the function of the other glutamatergic input, the layer 6 cortical one? Data presented below indicate that this cortical input acts as a modulator, much like classic modulator inputs, but with some differences detailed below. We refer to a glutamatergic input that functions mainly to carry information as a "driver" and one that serves to modulate as a "modulator" (Sherman and Guillery, 1998, 2013).

5.2 Properties of Glutamatergic Driver and Modulator Synapses

The distinction between the two types of glutamatergic input, drivers and modulators, has now been broadly applied to thalamus and cortex.

5.2.1 Basic Properties of Glutamatergic Drivers and Modulators

Glutamatergic inputs to neurons in thalamus and cortex can be classified into driver and modulator types based on a series of criterion parameters shown in Table 5.1 and Figure 5.2 (Sherman and Guillery, 1998, 2013).

- Glutamatergic driver inputs activate only ionotropic receptors, mainly AMPA and NMDA, whereas modulator inputs as well activate metabotropic glutamate receptors (Figure 5.2A). These latter can be either Group I, producing prolonged EPSPs, or Group II, producing prolonged IPSPs). It is important to emphasize the point that, with the activation of Group II metabotropic receptors, glutamate can act as an inhibitory neurotransmitter, and these receptors are found widely in cortex (Ohishi et al., 1993; Otani et al., 1999; Lee and Sherman, 2009a,b; Sherman, 2014; Venkatadri and Lee, 2014). The activation of metabotropic receptors by modulators is one of the most important functional distinctions between drivers

TABLE 5.1 Glutamatergic driver and modulator inputs

Criteria	Driver	Modulator
Criterion 1	Activates only ionotropic receptors	Activates ionotropic and metabotropic receptors
Criterion 2	Synapses show paired-pulse depression (high p)[a]	Synapses show paired-pulse facilitation (low p)
Criterion 3	Large initial EPSPs	Small initial EPSPs
Criterion 4	Little or no convergence onto target	Much convergence onto target
Criterion 5	Minority of glutamatergic inputs	Majority of glutamatergic inputs
Criterion 6	Large terminals on proximal dendrites	Small terminals on distal dendrites
Criterion 7	Thick axons with dense terminal arbors	Thin axons with delicate terminal arbors

[a]Probability of neurotransmitter release; see text for details.

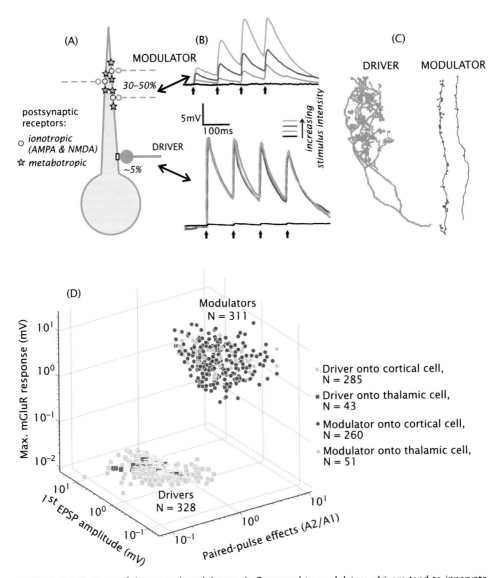

FIGURE 5.2 Features of drivers and modulators. A. Compared to modulators, drivers tend to innervate more proximal dendrites with larger diameter axons and larger terminals and with less convergence. They also produce many fewer synapses: for a thalamic relay cell, drivers produce about 5% and modulators, 30–50%, with the rest deriving from local GABAergic and brainstem modulatory inputs. Drivers also activate only ionotropic receptors, whereas modulators activate metabotropic receptors as well. B. Driver synapses are depressing and produce large initial EPSPs in an all-or-none fashion with increasing stimulation strength, implying relatively little convergence. In contrast, modulators produce initially small EPSPs that facilitate and are activated in a graded manner. C. Example of a driver versus modulator afferent. D. Three-dimensional scatterplot for inputs classified as driver or modulator to cells of thalamus and cortex; data from in vitro slice experiments in mice from one of the author's (SMS) laboratory. The three parameters are (1) the amplitude of the first EPSP elicited in a train at a stimulus level just above threshold; (2) a measure of paired-pulse effects (the amplitude of the second EPSP divided by the first; A2 divided by A1) for stimulus trains of 10–20 Hz; and (3) a measure of the response to synaptic activation of metabotropic glutamate receptors, taken as the maximum voltage deflection (i.e., depolarization or hyperpolarization) during the 300 msec postsynaptic response period to tetanic stimulation in the presence of AMPA and NMDA. Pathways tested here include various inputs to thalamus from cortex and subcortical sources, various thalamocortical pathways, and various intracortical pathways. D from Sherman, 2016.

and modulators, but recall that such activation requires modulator inputs to fire at moderately high rates (Viaene et al., 2013).

- Driver inputs show paired-pulse depression, implying a high average probability of neurotransmitter release, whereas modulators show paired-pulse facilitation, implying a low probability of release (Figure 5.2B,D). Note that this feature of modulators means that, at higher firing rates (i.e., >10 spikes/sec), such facilitation would mean even more neurotransmitter release, implying more will reach the more distant metabotropic receptors (see Figure 4.5).

- Driver inputs evoke a much larger initial EPSP than do modulator inputs (Figure 5.2B,D). This is presumably related to the higher probability of neurotransmitter release for driver inputs. It may also be related to evidence that driver synapses tend to be on more proximal dendrites, leading to less electrotonic decay of the EPSP from the synaptic location to the recording location in the soma (Figure 5.2A).

- Driver inputs tend to show less convergence onto target neurons than do modulator inputs (Figure 5.2A,B). For example, for geniculate relay cells of the cat, it has been estimated that retinal (driver) inputs have a much lower convergence rate (typically 2–5; see Cleland et al., 1971a,b; Usrey et al., 1999), than do layer 6 cortical (modulator) inputs (typically >100; Sherman and Koch, 1986). Evidence for such a convergence difference has also been seen for other circuits in thalamus and cortex (reviewed in Sherman and Guillery, 2013).

- In thalamic circuitry, synapses from modulator inputs to relay cells vastly outnumber those from driver inputs, with about 5% from drivers and 30–50% from glutamatergic modulators (Erisir et al., 1997b; Van Horn et al., 2000; Wang et al., 2002). It is striking that the functionally dominant input to the lateral geniculate nucleus, namely the retinal input, produces such a small minority of synapses, a point taken up again in Chapter 7. Such data from cortex are quite limited, but where they exist, there seems to be a similar quantitative relationship in numbers of driver versus modulator synapses. For example, geniculocortical synapses in layer 4 of cats and monkeys account for only 1–6% of synapses there (Peters and Payne, 1993; Peters et al., 1994; Ahmed et al., 1997).

- Driver inputs tend to have much larger terminals, on average, than do modulator inputs, both in thalamus and cortex (Sherman and Guillery, 2013). These differences seem quite pronounced in thalamus (Van Horn et al., 2000), but in cortex, it appears that modulatory terminals are all relatively small, whereas driver terminals have a wider size distribution, overlapping modulatory terminals at the small end and having a pronounced tail of larger terminals (Covic and Sherman, 2011; Viaene et al., 2011a,b; Petrof et al., 2015; Mo et al., 2017; Mo and Sherman, 2019).

- Where data are available, driver axons are larger diameter than are modulator axons (Figure 5.2C). For instance, retinal axons innervating geniculate relay cells are much larger diameter than are the innervating cortical layer 6 axons, but such data for cortical circuitry are generally lacking.

Figure 5.2D shows a three-dimensional scatterplot of differences between drivers and modulators based on recordings made from mouse brain slice preparations; here each data point represents a single cortical or thalamic neuron (keyed by color) for which an input identified as driver or modulator was activated. The axes represent three of the preceding criteria: the amplitude of the first EPSP evoked at just above activation threshold, the extent of paired-pulse effects, and the amplitude of any evoked metabotropic receptor response. Three conclusions can be drawn from this. (1) the classification is quite robust, since the scattering of points shows clear separation; (2) only two classes of glutamatergic input have so far been identified, although further study may identify more, especially beyond thalamus and cortex; and (3) properties of each type of input are quite similar in thalamus and cortex, although cortical driver inputs do show more variability, and it seems that these have subclasses (Viaene et al., 2011c).

5.2.2 Speculation on the Function of Glutamatergic Drivers and Modulators

While the driver/modulator classification seems clear for glutamatergic inputs in thalamus and cortex, this leaves open the question of its functional significance. The hypothesis we repeat here is that drivers carry the main information and modulators affect how driver inputs are processed (Sherman and Guillery, 1998, 2013; Sherman, 2014). One way to think about this with respect to sensory processing is that driver inputs provide the basic receptive field features to their postsynaptic cells, whereas modulators do something else, which we suggest is modulation of the processing of driver inputs in various thalamic and cortical circuits.

5.2.2.1 Driver Function

Once again, a consideration of the lateral geniculate nucleus helps to explain the function of drivers. Consider the functional significance of retinal versus cortical layer 6 input to geniculate relay cells (Figure 5.1): retinal (driver) input provides the geniculate neuron with its receptive field properties, whereas the cortical input clearly does not. Similar evidence exists for driver inputs to the somatosensory and auditory thalamic nuclei (reviewed in Sherman and Guillery, 2013).

That action of driver inputs in thalamus seems relatively straightforward is due partly to the lack of convergence of these inputs reflected in the relative lack of elaboration of receptive field properties between retinal axons and their geniculate relay cell targets. Whether this lack of integrative action is a general feature of thalamus is considered further in Chapter 7. However, when we consider cortex, this hypothesis for driver functioning rests on less data and remains more speculative. For example, geniculocortical input to layer 4 of visual cortex, which has the properties of a driver (Lee and Sherman, 2008; Viaene et al., 2011a), also imparts receptive field properties to its target cortical cells (Hubel and Wiesel, 1962; Ferster, 1987; Ferster et al., 1996; Alonso et al., 2001). However, in this case there is receptive field elaboration due to the convergence of geniculate inputs to layer 4 cells. Other examples of driver convergence in visual cortex is the establishment of binocular receptive fields somewhere in cortical circuitry from geniculate inputs that are monocular, and there is evidence that complex receptive fields are based on convergence from simple cell afferents (Hubel and Wiesel, 1962; Alonso and Martinez, 1998). It seems plausible that other examples of convergence of driver inputs carrying different information and possibly from different areas converge onto single cells that reflect cortical processing and integration of such information, but such examples remain to be documented.

5.2.2.2 Modulator Function

For a variety of reasons, modulator inputs seem poorly organized to be efficient carriers of information. Their facilitating synapses produce weak initial EPSPs that grow only after a sustained train of action potentials fired at greater than 10 spikes/sec, which means that any information transferred across the synapse would suffer poor transmission early in such a train and successful transmission only much later as the EPSPs become sufficiently large to influence spiking in the postsynaptic cell. This means that early information conveyed by the train is seriously compromised. Also, the large convergence of postsynaptic modulatory responses in their inputs suggests that even to drive postsynaptic targets would require synchronous firing that seems an unlikely common natural occurrence. Finally, the need for higher firing rates to achieve facilitation also means that metabotropic receptors will be activated, and this poses additional timing problems. The prolonged EPSP means that there cannot be a one-to-one matching between EPSPs and input action potentials, and so the temporal patterning of the input is further compromised. Furthermore, the response outlasts activity in the input, often by seconds (Govindaiah and Cox, 2004), which may be useful for modulation but further distorts the temporal transmission of information.

Instead of being organized chiefly to transmit information in circuits, like driver glutamatergic inputs, we argue that modulators modulate much like their classic counterparts (i.e., cholinergic, serotonergic, etc.). Common to these functions is the activation of metabotropic receptors. Evidence for modulation by metabotropic glutamate receptors derives mostly from work in slices of mouse brain. Such action takes two main forms based on whether the receptors are postsynaptic or on presynaptic terminals. Postsynaptic action of these receptors produces sustained PSPs that can affect overall excitability of the target cell and also control the inactivation states of voltage- and time-gated channels like the T-type Ca^{2+} channel as described in Chapter 3 (reviewed in Sherman and Guillery, 2013). Note that this activation includes sustained EPSPs as well as IPSPs (Sherman and Guillery, 2013). In some experiments, activation of presynaptic metabotropic receptors are mostly Group II, and activation of these on glutamatergic driver inputs reduces the evoked EPSPs on target cells, presumably by hyperpolarizing the terminals and thus reducing evoked neurotransmitter release (DePasquale and Sherman, 2012, 2013).

One might ask: Given the plethora of classical modulatory inputs to thalamus and cortex, why also have glutamatergic modulators? As noted, classical modulators tend to be diffusely organized (see Figure 4.3) and seem more relevant to overall behavioral state, whereas glutamatergic inputs, including glutamatergic modulators, are generally highly topographic. Furthermore, the classic modulatory pathways, which derive from brainstem structures, necessarily reflect only brainstem processing, whereas, the glutamatergic modulators discussed here benefit from processing in thalamus and cortex. Such topographic modulation controlled by cortex and thalamus is needed for cognitive processes that require localized effects, such as focal or covert attention, adaptation, learning and memory, etc.

5.2.2.3 Drivers and Modulators in the Lateral Geniculate Nucleus

Consider the circuitry of the lateral geniculate nucleus, with its two main glutamatergic inputs from retina and layer 6 of visual cortex (see above discussion). Clearly, the retinal input is the critical one for the relay function of geniculate neurons, yet this input provides only about 5% of the synaptic input to these neurons, whereas the cortical input provides roughly an order of magnitude more. Using the "anatomical democracy" approach, one would likely conclude that the numerically dominant cortical input provides the critical information to be relayed (back to itself!), while the retinal input is a minor player of obscure function. We have known for decades that the retinal input is the true driver here, so, to our knowledge, there has been no such embarrassing suggestion.

Nonetheless, there remains a tendency to treat other glutamatergic pathways, such as projections between cortical areas, as a homogeneous entity of importance related to its size. The preceding example of circuitry of the lateral geniculate nucleus serves as a reason to question this practice and highlights the importance of identifying drivers and modulators in glutamatergic circuits. Other examples of the value of this classification are described in Chapter 6.

5.2.3 Examples of Identified Driver and Modulator Inputs

Examples of glutamatergic inputs to thalamic or cortical neurons so far identified as driver or modulator are listed in Table 5.2. The actual data are summarized in Sherman and Guillery (2013; see also Petrof et al., 2015; Mo et al., 2017; Mo and Sherman, 2019). If we consider

TABLE 5.2 Identified Glutamatergic Driver & Modulator Inputs in Mouse Brain Slices

	Onto Thalamic Neurons
	Onto Cortical Neurons

DRIVERS	*MODULATORS*
retina to LGN; medial lemniscus to VPm; inferior colliculus (core) to MGNv; mammillary body to AD	layer 6 to LGN, VPm, MGNv, Pul, POm, and MGNd
cortical layer 5 to POm, Pul, MGNd, VA/VL	inferior colliculus (shell) to MGNd
a minority from spinal nucleus of the fifth nerve to POm	a majority from spinal nucleus of the fifth nerve to POm a majority from VPm to S1 and MGNv to A1, layers 2/3
VPm to S1 and MGNv to A1, layers 2-6 and minority to layers 2/3	POm to S1, and MGND to A1, layers 2-6
many interareal cortical projections from V1 to V2, V2 to V1, A1 to A2, and A2 to A1; all from S1 to M1	layer 6 to layer 4 layer 2/3 to layer 2/3
layer 4 to layer 2/3	many interareal cortical projections from V1 to V2, V2 to V1, A1 to A2, and A2 to A1

Abbreviations: *A1 & A2,* primary and secondary auditory cortex; *AD,* anterior dorsal nucleus; *LGN,* lateral geniculate nucleus; *M1,* primary motor cortex; *MGNd & MGNv;* dorsal and ventral divisions of medial geniculate nucleus; *POm,* posterior medial nucleus; *Pul,* pulvinar; *S1 & S2,* primary and secondary somatosensory cortex; *V1 & V2;* primary and secondary visual cortex; *VA/VL,* ventral lateral and anterior thalamic nuclei; *VPm,* ventral posterior medial nucleus. Data from laboratory of SMS.

the example of interareal corticocortical connections, these include both driver and modulator classes, but their distribution depends on the origin and/or target layers (Covic and Sherman, 2011; DePasquale and Sherman, 2011). This plus the fact that the importance of a corticocortical connection is still often determined by its size emphasizes the point that identifying driver versus modulator components of these connections is a necessary prerequisite to understanding their functioning. This is especially true if relative numbers of driver and modulator pathways in cortex are anything like those in thalamus, where the driver inputs are vastly outnumbered by modulator inputs, which would mean that lumping them together in cortical circuitry would mask the ability to follow processing of information in cortex.

5.2.4 Possible Differences in Properties at Different Terminals of an Axon

Do all synaptic terminals from a glutamatergic axon need to be of the same driver or modulator type? For instance, driver inputs to thalamus often or always involve branching axons that innervate extrathalamic targets as well. Do these extrathalamic targets also receive driver inputs from these branches? Branching axons are a ubiquitous feature of the central nervous system, and the topic of branching glutamatergic axons is taken up in more detail in Chapter 13.

An elegant study by Reyes et al. (1998) offers insight into this issue. These authors intracellularly recorded simultaneously from several cells in cortex with the arrangement that one of the recorded cells was found to be presynaptic to two or more of the others, and all of the connections here were glutamatergic. It was found that the afferent cell could produce paired-pulse depression (a driver feature) at one target cell and paired-pulse facilitation (a modulator feature) at another. This demonstrates that different branches of an afferent axon can quite plausibly evoke different synaptic properties at its different targets.

5.3 Summary and Concluding Remarks

As emphasized at the beginning of this chapter, classification of constitutive elements is a critical early step in exploring complex systems, such as the brain. Until recently, classification of circuits has largely been limited to identifying the transmitter(s) used by different afferent groupings. However, such classification can be usefully extended further by identifying transmitter subclasses, just as identifying subclasses of retinal ganglion cells has provided important insights into the functional organization of visual pathways.

We have shown that the main excitatory transmitter used by brain circuits, namely glutamate, is used by at least two very different classes of input, which are called drivers and modulators. These classes, which are clearly present in thalamus and cortex, can be identified on the basis of multiple criteria (Table 5.1 and Figure 5.2). Glutamatergic drivers, we argue, carry the main information for further processing from neuron to neuron; glutamatergic modulators modulate much as do classical modulatory systems (e.g., cholinergic, serotonergic, etc.) with the added advantage of being topographically organized and based on thalamic and cortical processing.

Further glutamatergic classes and subclasses may still be found. We argue that identifying which type of glutamatergic input is involved in any circuit is necessary for understanding the functional organization of that circuit and following the flow of information as it is processed by the brain. The value of subclassification of glutamatergic afferents seems clear. Perhaps subclassification of other afferent transmitter systems (e.g., cholinergic, GABAergic) would also prove useful.

5.4 Some Outstanding Questions

1. Can a glutamatergic afferent act as a driver under some conditions and as a modulator under others?
2. What are the classes of glutamatergic inputs in parts of the brain other than thalamus and cortex, or, do classes of glutamatergic input exist other than drivers and modulators?
3. Do all driver inputs in the brain use glutamate as a transmitter and activate only ionotropic glutamate receptors, or, for instance, might some be cholinergic, activating nicotinic receptors?
4. Do modulators always outnumber drivers in all parts of the brain?
5. Must all synaptic terminals within a single arbor of one axon have the same effect on their target as regards driver or modulator function?
6. What other differences exist between driver and modulator synapses? For instance, do both show the same sorts of plasticity, such as long-term potentiation or depression; do both have similar or different patterns of presynaptic receptors on their terminals; etc.?
7. What effects beyond opening or closing of ion channels does activation of metabotropic receptors produce?
8. In activation of some modulator pathways, both ionotropic and metabotropic glutamate receptors are activated, but do individual axons activate both, or are these separate populations? Since a given modulatory input can both excite and inhibit by activating multiple receptor types, do individual inputs just excite (by activating ionotropic and/or Group I metabotropic receptors) and others just inhibit (by activating Group II metabotropic receptors)?

6

First and Higher Order Thalamic Relays

The function of a thalamic relay is largely defined by its driver input. For example, the lateral geniculate nucleus can be defined as a relay of retinal input to cortex; the ventral posterior nucleus, as a relay of medial lemniscal input, etc. These primary sensory thalamic nuclei have long been understood to relay subcortical information from retina, spinal cord, or brainstem to cortex. More recently, driver inputs to many other thalamic nuclei were identified as originating in layer 5 of cortex, and this has led to a classification of thalamic relays into two types: *first order relays* have subcortical driver inputs, whereas *higher order relays* are driven by layer 5 of cortex (Sherman and Guillery, 2001; Sherman, 2013, 2016).

6.1 Evidence That Layer 5 Corticothalamic Inputs Are Driver

In the previous chapter, we presented evidence that the layer 6 corticothalamic pathway was largely or exclusively organized in a feedback manner and functionally acts as a modulatory input. Layer 5 corticothalamic inputs have been known to exist for decades, but only relatively recently has there been evidence to distinguish them functionally from the layer 6 projections (reviewed in Guillery, 1995; Sherman and Guillery, 2013; Usrey and Sherman, 2019).

6.1.1 Anatomical Evidence

The first evidence that distinguished layer 5 from layer 6 corticothalamic terminals came from morphological studies using the electron microscope (reviewed in Guillery, 1995; Sherman and Guillery, 2013). These studies showed that such corticothalamic terminals labeled without regard to laminar origin and found in certain thalamic nuclei, which we now regard as first order, are rather uniformly small and end on distal dendrites. In contrast, such terminals found in other thalamic nuclei, which we now regard as higher order, include the smaller, more distal terminals but also include larger ones found more proximally. The arrival of modern tracing techniques showing the relationship between terminal morphology and cortical layer of origin allowed the determination that the smaller, distal terminals originate from layer 6 and the larger, proximal ones, from layer 5 (Guillery, 1969a,b; Wilson et al., 1984; Bourassa et al., 1995; Bourassa and Deschênes, 1995; Ojima et al., 1996; Rockland, 1996; Guillery et al., 2001; Kakei et al., 2001; Sherman and Guillery, 2013). Thus, all thalamic nuclei receive a layer 6 corticothalamic projection, but only a subset in addition receive a layer 5 projection.

The observation that some thalamic nuclei receive a cortical layer 5 input and others do not is sufficient evidence to distinguish the two types of thalamic relay that we refer to as first and higher order. That is, the corticothalamic layer 5 projection to thalamus defines those targets as higher order thalamic relays, and the other thalamic relays are thus first order. Given that these data include observations from rodents, carnivores, insectivores, and primates (reviewed in Sherman and Guillery, 2013), they can be regarded as reflecting a general mammalian plan for thalamic relays, although the details for brains that have relatively few functionally distinct cortical areas have not been defined, and these could serve to provide useful clues to the evolutionary origins of these connections.

More recent evidence suggests that it is a general feature of cortex to have a layer 5 projection to thalamus, meaning that nearly every area—and perhaps every one—does so, and so far, no exception to this pattern has been described (Sherman and Guillery, 2013; Prasad et al., 2020).

6.1.2 Physiological Evidence

The first physiological demonstration that there are two functional types of synaptic input, presumed to be of cortical origin, to a higher order nucleus was based on recordings from the pulvinar of the rat (Li et al., 2003a): one type of input showed paired-pulse depression (a driver characteristic) and the other, paired-pulse facilitation (a modulator characteristic). Soon after, these differences were attributed to differences between inputs from different cortical layers in a study of the input from the primary somatosensory cortex to the posterior medial nucleus of mice: the layer 6 input was identified as a modulator, and the layer 5, as a driver (Reichova and Sherman, 2004). This difference has generally held up for layer 6 inputs to all thalamic nuclei and layer 5 inputs to higher order relays (reviewed in (Sherman and Guillery, 2013).

6.2 Higher Order Thalamic Relays as Links in Transthalamic Corticocortical Communication

Given that higher order relays project to cortex and receive driving input from layer 5 of cortex, it follows that these relays are central components of transthalamic, corticocortical circuits (Sherman and Guillery, 2001, 2013; Sherman, 2016). This appreciation of the existence of higher order relays provides a straightforward hypothesis for much of thalamus that has heretofore been rather mysterious functionally: these higher order relays, which we estimate to be the majority of thalamus by volume, are involved in the transfer of information between cortical areas.

One generalization of a pattern difference between corticothalamic layer 5 and 6 projections is that the latter tend to be organized mostly in a feedback manner, meaning that the layer 6 target is the same thalamic region from which the thalamocortical projection arises to that cortical area, whereas the former is usually not a feedback pattern, meaning that the corticothalamic target is a thalamic zone innervating a region other than the cortical region of the layer 5 afferent cells (Van Horn and Sherman, 2004; Llano and Sherman, 2008). Some exceptions to these patterns are noted below.

Given that most or all cortical areas have a layer 5 projection to thalamus (which, by definition, means that these thalamic targets are higher order relays), and available data indicate that these layer 5 inputs are driver, it seems pretty clear that transthalamic pathways are a ubiquitous feature of cortical processing. In turn, this means that a much better understanding of the nature and diversity in these pathways is an absolute necessity for a general understanding of the functional organization of cortex.

Other, scattered evidence for the existence and importance of transthalamic circuitry exists. Suppression of primary somatosensory cortex eliminates layer 6 input to cells of the ventral posterior medial nucleus but has little effect on the basic receptive field structure of these cells, whereas such suppression also eliminates layer 5 input to cells of the posterior medial nucleus, and this procedure renders these latter thalamic cells insensitive to sensory stimulation (Diamond et al., 1992). This finding is in agreement with the conclusion that the posterior medial nucleus receives driver input from the primary somatosensory cortex.

Less comprehensive but nonetheless compelling evidence also exists. For instance, the pulvinar in both cat and monkey has neurons that seem to possess receptive field properties inherited from cortical input, suggesting that they receive cortical driver input and serve as a link in transthalamic circuitry (Chalupa et al., 1972; Bender, 1983; Merabet et al., 1998; Zhou et al., 2016). Moreover, other evidence indicates that the pulvinar in the monkey participates in transthalamic pathways to regulate information transfer between cortical areas (Saalmann et al., 2012). Studies in mice indicate that pulvinar neurons that project to higher visual cortical areas derive information to be relayed from primary visual cortex (Blot et al., 2021). Finally, a study using optogenetics and Ca^{2+} imaging in mice has shown that activation of cortical layer 5 cells produces waves of activity in other cortical areas, and these activation waves depend on transthalamic pathways (Stroh et al., 2013).

6.2.1 Feedforward Transthalamic Pathways

The organization of higher order relays rather obviously has led to the hypothesis that they are a fundamental route for corticocortical communication. Figure 6.1 illustrates this point, starting with the visual system as an example (Figure 6.1A). Although there are a number of subtle differences between first and higher order thalamic nuclei, differences that are elaborated in Chapter 7, here these are ignored. To a first approximation, the main difference between these thalamic types is the source of the driver input: subcortical for first order relays (e.g., retina for the lateral geniculate nucleus) and cortical layer 5 for higher order relays (e.g., from visual cortex to pulvinar).

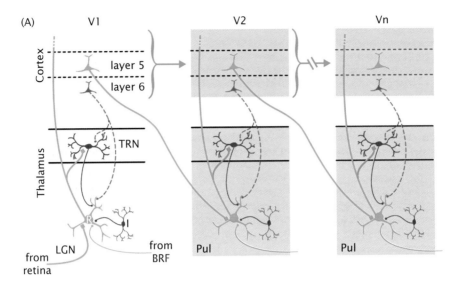

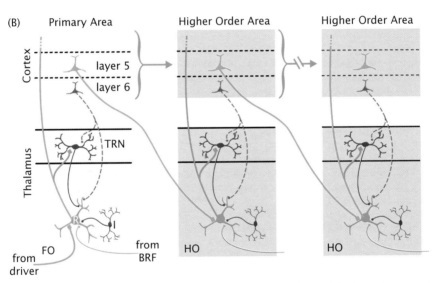

FIGURE 6.1 Schematic diagrams showing organizational features of first and higher order thalamic nuclei. A. Thalamocortical arrangements in the visual system. A first order nucleus (FO; LGN) represents the first relay of visual information to V1 or primary visual cortex area. A higher order nucleus (HO; Pul) relays information from layer 5 of one visual cortical area to another. This relay can be between primary (V1) and secondary (V2) cortical areas as shown or between two higher order visual cortical areas. The important difference between them is the driver input, which is subcortical (retinal) for a first order thalamic nucleus and from layer 5 of cortex for a higher order one. Note that all thalamic nuclei receive an input from layer 6 of cortex, which is modulatory and mostly feedback, but higher order nuclei in addition receive a layer 5 input from cortex, which in these examples is feedforward. B. More general thalamocortical arrangements.

Abbreviations: *BRF*, brainstem reticular formation; *FO*, first order; *HO*, higher order; *I*, interneuron; *LGN*, lateral geniculate nucleus; *R*, relay cell; *TRN*, thalamic reticular nucleus. Redrawn from Sherman, 2007.

Note that Figure 6.1A shows the transthalamic pathways organized in a feedforward manner climbing the hierarchy of visual cortical areas (Van Essen et al., 1992; Coogan and Burkhalter, 1993; Scannell et al., 1995). (Feedback configurations are considered below.) In Figure 6.1A, the first order relay is the lateral geniculate nucleus and the higher order relay is the pulvinar. The pulvinar innervates all the areas of visual cortex, which in monkeys and humans involves 30–40 separate areas and includes nearly the entire back half of the cortical mantle. The hypothesis is that these layer 5 transthalamic pathways involving pulvinar include not just those from the primary to secondary visual areas, but also are involved in similar relays as the cortical hierarchy is ascended, although specific examples of this for higher order cortical areas remains to be acquired. The pulvinar is the largest thalamic nucleus, being 30–50 times larger in volume than the lateral geniculate nucleus in monkeys and humans, so it would seem that it has plenty of bandwidth for such processing.

Figure 6.1A also shows direct connections between the same cortical areas, indicating a pattern whereby these cortical areas are connected both directly and transthalamically by parallel pathways. This raises three key questions for which we have as yet no clear answers: How often are two cortical areas connected by both pathways in parallel, or, how often are two cortical areas connected by only a direct or transthalamic circuit? What is different in the information content carried by the direct versus transthalamic pathway? Why is one information stream filtered through the thalamus? A partial answer to this last question is that such information transferred by a transthalamic pathway can be modulated or gated by thalamic circuitry in ways unavailable to direct corticocortical pathways.

Figure 6.1B makes the point that this arrangement is not just limited to the visual system but also has been extended to other examples as well. This includes the somatosensory and auditory pathways (reviewed in Sherman and Guillery, 2013): in somatosensation, the ventral posterior nucleus is first order, and the posterior medial nucleus, higher order; in audition, the ventral division of the medial geniculate nucleus is first order, and the dorsal division, higher order. Recently, a transthalamic sensorimotor circuit has been identified from the primary somatosensory to the primary motor cortex via the posterior medial nucleus, which extends such circuitry beyond purely sensory areas of cortex (Mo and Sherman, 2019). In each of these examples, both a direct and transthalamic pathway was identified, which provides a partial answer to the first question raised, but clearly many more examples are needed. It is likely that transthalamic circuitry may be a common feature of corticothalamic processing, because layer 5 inputs to thalamus have been identified from many areas of cortex (Clasca et al., 1995; Cappe et al., 2007; Xiao et al., 2009; Kita and Kita, 2012; Mitchell, 2015; Economo et al., 2018; Prasad et al., 2020; Winnubst et al., 2019), and, as noted above, it is plausible that every cortical area has a layer 5 projection to thalamus.

Limited data suggest that these feedforward transthalamic circuits are comprised of driver inputs, which in turn suggests that they are information-bearing. So far, all layer 5 corticothalamic inputs to thalamus have shown driver characteristics (Sherman and Guillery, 2013; Miller and Sherman, 2019; Mo and Sherman, 2019), and the few examples of feedforward higher order thalamocortical inputs are also driver. But the examples are very few indeed, are limited to the mouse, and include just transthalamic circuits from the primary to the secondary somatosensory cortex (via the posterior medial nucleus), from the primary to

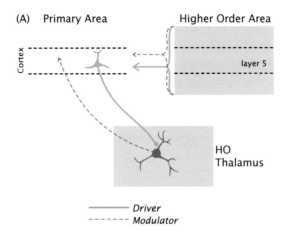

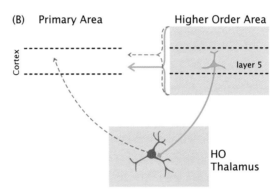

FIGURE 6.2 Feedback direct and transthalamic corticocortical circuits to primary cortical area. The direct circuits contain both drivers and modulators (Covic and Sherman, 2011; DePasquale and Sherman, 2011). The transthalamic circuits involve a layer 5 driver input to the higher order thalamic relay, and the thalamocortical input is modulatory (Miller and Sherman, 2019). A. Transthalamic feedback from primary cortex to itself. B. Transthalamic feedback from higher order cortical area to primary cortex.

the secondary auditory cortex (via the dorsal division of the medial geniculate nucleus), and from the primary somatosensory cortex to the primary motor cortex (via the posterior medial nucleus; Lee and Sherman, 2008; Theyel et al., 2010; Viaene et al., 2011a; Sherman and Guillery, 2013; Miller and Sherman, 2019; Mo and Sherman, 2019). Clearly, more examples are needed both in terms of more thalamocortical regions as well as from more species.

6.2.2 Feedback Transthalamic Pathways

Even more limited data are available for feedback transthalamic pathways. At present we can say little more than they seem to exist, but how common they are and what sort of functions they subserve are matters pretty much limited to speculation. It is somewhat confusing that there seem to be two different types of transthalamic circuitry that can both be considered "feedback." Both are shown in Figure 6.2.

Figure 6.2A shows a transthalamic feedback circuit in which the thalamic relay is back to the same area from which its layer 5 input arises. This has been demonstrated in the mouse for a circuit from the primary visual cortex to the pulvinar back to the primary visual cortex as well as a circuit from the primary somatosensory cortex to posterior medial nucleus back to the primary somatosensory cortex (Miller and Sherman, 2019). Figure 6.2B shows a feedback circuit from a higher cortical area through thalamus to a primary area. This has been demonstrated in the mouse for the secondary to primary visual and somatosensory cortices. What is striking in both cases is the finding that the projection from the higher order thalamus to primary cortical areas is purely modulatory, which is in stark contrast to feedforward projections from higher order thalamus to higher cortical areas, which is chiefly or exclusively driver (Lee and Sherman, 2008; Viaene et al., 2011; Miller and Sherman, 2019; Mo and Sherman, 2019).

Note also in the example shown in Figure 6.2B that there is a parallel organization of direct and transthalamic feedback circuits from higher to lower cortical areas. Thus the parallel organization may apply to both feedforward and feedback corticocortical pathways. The same three questions arise for feedback circuitry as for feedforward circuitry, and these are worth repeating here with minor modifications: How often are two cortical areas connected by both pathways in parallel, or, how often are two cortical areas connected by only a direct or transthalamic circuit? What is different in the messages carried by the direct versus the transthalamic pathway? Why is one pathway filtered through the thalamus?

The implication is that, whereas feedforward transthalamic pathways are mainly or exclusively driver, the feedback ones seem mostly modulatory. A proviso for this last point is a recent study showing a feedback transthalamic circuit in the mouse similar to that in Figure 6.2A between the anterolateral motor cortex and ventral medial thalamus in which the thalamocortical limb appears to be strong enough to be characterized as driver, although that property was not explicitly tested (Guo et al., 2018).

Note that, like the feedforward corticocortical circuits, the feedback circuits in Figure 6.2B involve both transthalamic and direct moieties. The transthalamic circuit provides purely modulatory input to the primary cortical areas, whereas the direct circuit includes both driver and modulator components (Covic and Sherman, 2011; DePasquale and Sherman, 2011).

6.3 Examples of Value of Driver/Modulator and First/Higher Order Classifications

The older idea that glutamatergic inputs represent a somewhat homogeneous functional class organized to convey information to postsynaptic targets in turn led to the idea that the importance of a glutamatergic pathway was related to its size, as if these circuits acted in a sort of anatomical democracy. We have already shown in Chapter 5 how a failure to appreciate the driver/modulator distinction for geniculate circuitry could provide a misleading interpretation of the functional organization of the lateral geniculate nucleus. Three additional examples provided here indicate the value of the driver/modulator identification over this older view, and each further involves the identification of transthalamic pathways.

6.3.1 Ascending Auditory Pathways Through Thalamus

The first of these examples is illustrated in Figure 6.3A,B, showing the relay of auditory information through thalamus to cortex. The classical view (Figure 6.3A) is that two independent pathways from the inferior colliculus through the medial geniculate nucleus represent parallel streams of auditory information to auditory cortex (Hu, 2003; Kraus and Nicol, 2005). Specifically, the lemniscal stream arises from the core region of the inferior colliculus and is relayed by the ventral division of the medial geniculate nucleus to the primary auditory cortex; the paralemniscal stream arises from the shell region of the inferior colliculus and is relayed by the dorsal division of the medial geniculate nucleus to the secondary auditory cortex. However, an analysis of these circuits based on identifying drivers versus modulators suggests a very different functional organization (Figure 6.3B; Lee and Sherman, 2010). This analysis shows that the lemniscal pathway is indeed comprised of drivers, and so this pathway to the primary auditory cortex seems to be an information route. However, the paralemniscal input from the shell region of the inferior colliculus to the dorsal division of the medial geniculate nucleus is entirely modulator. The actual driver for these latter medial geniculate neurons arises from layer 5 of the primary auditory cortex, representing a transthalamic information route from the primary to secondary auditory cortex via the dorsal division of the medial geniculate. It seems that what used to be regarded as a paralemniscal information input to thalamus instead acts to modulate or gate activity in the transthalamic circuit. Thus, the different functional organization of these ascending auditory pathways shown in Figure 6.3A,B illustrates one example of the value of the driver/modulator and first order/higher order classifications.

6.3.2 Ascending Somatosensory Pathways Through Thalamus

Another closely related example exists for the somatosensory system, for which parallel lemniscal and paralemniscal information routes have also been claimed (Figure 6.3C; Ahissar et al., 2000; Yu et al., 2006; Nakamura et al., 2009; Sitnikova and Raevskii, 2010): the lemniscal, from the principal nucleus of the fifth nerve to the ventral posterior nucleus to the primary somatosensory cortex, and the paralemniscal, from the spinal nucleus of the fifth nerve to the posterior medial nucleus to secondary somatosensory cortex. Again, this implies that the paralemniscal input is comprised of driver components, but it has recently been demonstrated that most of this input is modulatory (Figure 6.3D; Mo et al., 2017), suggesting again that the main purpose of this paralemniscal input to thalamus is to modulate or gate the transthalamic pathway.

6.3.3 Ascending Visual Pathways Through Thalamus

The equivalent visual pathways to those just described for the auditory and somatosensory systems involve the superior colliculus, which innervates both the lateral geniculate nucleus and pulvinar (Sherman and Guillery, 2013, 2014). The notion is that a pathway parallel to the retino-geniculo-cortical pathway involves a retino-colliculo-thalamo-cortical one. At

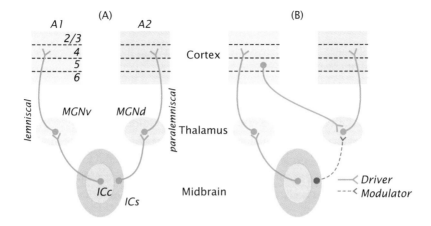

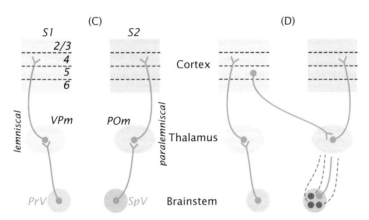

FIGURE 6.3 Differing views of lemniscal and paralemniscal paths through thalamus to cortex. A. Conventional view for auditory pathways that shows two parallel information routes to cortex, a lemniscal stream from the core region of the inferior colliculus (ICc) to the ventral division of the medial geniculate nucleus (MGNv) to primary auditory cortex (A1), and paralemniscal stream from the shell region of the inferior colliculus (ICs) to the dorsal division of the medial geniculate nucleus (MGNd) to secondary auditory cortex (A2). B. More recent view (Lee and Sherman, 2010). Whereas the lemniscal stream through the ventral division of the medial geniculate nucleus is comprised of driver inputs and thus seems to be an information route, the paralemniscal input to the dorsal division of the medial geniculate nucleus is modulator and thus seems to operate not as an information stream but rather to modulate the transthalamic circuit from primary to secondary auditory cortex. C,D. Similar differing views for the somatosensory pathways. C. In the conventional view, the lemniscal stream via the principal nucleus of the fifth nerve, PrV) through the ventral posteromedial nucleus (VPm) to primary somatosensory cortex and paralemniscal stream via the spinal nucleus of the fifth nerve, SpV) through the posterior medial nucleus (POm) to secondary somatosensory cortex are organized as parallel information routes. D. More recent view (Mo et al., 2017). Whereas the lemniscal stream through the ventral posteromedial nucleus is comprised of driver inputs and thus seems to be an information route, the paralemniscal input to the posterior medial nucleus is mostly modulator and thus seems chiefly to operate not as an information stream but rather to modulate the transthalamic circuit from primary to secondary somatosensory cortex.

issue in the context of this account is the nature of the colliculothalamic inputs: driver or modulator.

Regarding the lateral geniculate nucleus, the superior colliculus mostly innervates the C layers in the cat, the koniocellular layers in the primate, and the shell region in the mouse. We argue in Chapter 11 that the W and K pathways are homologous, and evidence from the mouse further suggests that the shell region is homologous to the geniculate W and K layers in the cat and primate, respectively (Kerschensteiner and Guido, 2017; Guido, 2018). No evidence yet exists for the driver/modulator identity of the colliculogeniculate input in the cat or primate. However, recent evidence suggests that the collicular input to the shell geniculate cells in the mouse is driver and, furthermore, that these geniculate cells also receive convergent retinal input (Bickford et al., 2015). Thus, there is a plausible basis for a retino-colliculo-geniculo-cortical information route.

As for the colliculopulvinar input, there have been numerous anatomical and physiological descriptions of the pathway in several species, but none has yet clearly identified the afferent class of glutamatergic input, driver, modulator, or other (Robson and Hall, 1977; Stepniewska et al., 1999, 2000; Kelly et al., 2003; Chomsung et al., 2008; Masterson et al., 2009; Wei et al., 2011; Baldwin et al., 2013, 2017), although an anatomical description of large colliculopulvinar terminals in the cat (Kelly et al., 2003) suggests that at least some of this input may be driver. Furthermore, a recent study in the mouse indicates that there is a pathway from the colliculus to the postrhinal cortex, a higher order visual area, and responses in this area survive removal of V1 but not the superior colliculus (Beltramo and Scanziani, 2019). This indicates an information-bearing colliculo-thalamo-cortical pathway organized in parallel to the geniculocortical. However, the thalamic nucleus or nuclei involved in this parallel circuit remain undefined, since the superior colliculus projects to the pulvinar, lateral geniculate nucleus, and the medial dorsal nucleus, among other possible targets.

6.3.4 Basal Ganglia Innervation of Thalamus

A final example, regarding basal ganglia and its relationship to thalamus and cortex, is shown in Figure 6.4A,B. Figure 6.4A shows the textbook view that still prevails (Kandel et al., 2000; McHaffie et al., 2005; Purves et al., 2012). That is, a loop of information flow exists based on projections from cortex to the basal ganglia to thalamus and back to cortex. But again, a consideration of actual information flow carried by drivers changes how this circuitry may be interpreted. The first problem with the textbook view is that the projection from the basal ganglia to thalamus is entirely GABAergic and powerfully inhibits its target thalamic neurons (MacLeod et al., 1980; Tanibuchi et al., 2009; Goldberg and Fee, 2012; Kim et al., 2017). GABAergic synapses seem a poor substrate for information transfer (Smith and Sherman, 2002) and appear instead designed to control overall excitability and gain in circuits. Thus the loop of efficient information flow suggested in Figure 6.4A would be interrupted in the GABAergic projection from basal ganglia to thalamus.

Figure 6.4B shows a very different interpretation of this circuitry. The main thalamic nuclei involved in this circuit are the ventral anterior and lateral nuclei, which relay motor information to motor cortical areas. These nuclei are organized more like a mosaic, and evidence exists in the monkey (Sakai et al., 1996) and rat (Kuramoto et al., 2011) that the

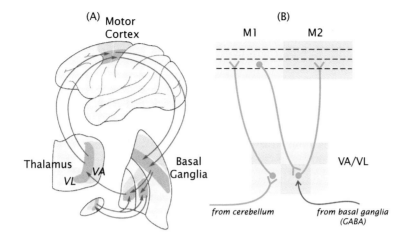

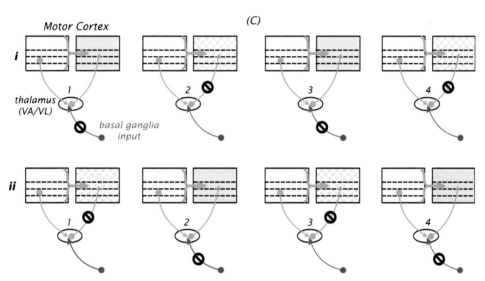

FIGURE 6.4 Differing views of relationship of basal ganglia with thalamocortical processing. A. Textbook view showing an information loop from cortex to basal ganglia to thalamus to cortex, and so on (redrawn from (Kandel et al., 2000). B. View based on driver/modulator framework. The inputs from basal ganglia and cerebellum to the motor thalamus (the ventral anterior/ventral lateral complex, or VA/VL) target mainly separate zones. A layer 5 projection from motor cortex to the motor thalamus exists, but it is not clear which zone(s) it innervates. One suggestion is that at least some of the layer 5 input represents the start of a transthalamic pathway (between primary and secondary motor cortices, M1 and M2, in this example) and that the basal ganglia input, which is GABAergic, gates this circuit. See text for further details. C. Examples of how this gating would work to determine which cortical areas are connected via transthalamic pathways. If the basal ganglia input is active, it inhibits the thalamic relay cell, shutting down the transthalamic pathway, but if the basal ganglia input is inactive, it allows functioning of the transthalamic pathway. Ci. In these examples, basal ganglia input is active for thalamic cell groups 2 and 4 but not 1 and 3. Thus the cortical areas related to thalamic groups 1 and 3 are connected both via their direct pathways *and* transthalamically, whereas those areas related to thalamic groups 2 and 4 are connected only via their direct pathways. Cii. A different pattern of basal ganglia activity leads to a different pattern of cortical areas connected only via direct projections and those also connected via transthalamic pathways.

inputs to this thalamic mosaic from the cerebellum and from the basal ganglia occupy different territories. Furthermore, cortical layer 5 of motor cortex projects to this mosaic complex forming large, localized terminals there (McFarland and Haber, 2002; Kultas-Ilinsky et al., 2003; Prasad et al., 2020), which suggests a driver input and thus defines a higher order thalamic zone of transthalamic processing much like that described in Figure 6.3B,C. We do not yet know which zone(s) the layer 5 projection innervates in the mosaic of ventral anterior and ventral lateral nuclei, but it seems likely that at least some of the projection is aimed at the zone innervated by the basal ganglia. If so, then the input from the basal ganglia serves mainly to modulate or gate the layer 5 input to the transthalamic pathways involving the ventral anterior and ventral lateral anterior nuclei. This represents a very different view than standard textbook accounts of the function of the basal ganglia and their influence on motor cortex.

Figure 6.4C further elaborates on this idea for the basal ganglia input to thalamus. Imagine four examples of corticocortical connections among motor cortical areas that include both direct and transthalamic pathways. The direct connections may always be active but communication via the transthalamic circuit is gated by the basal ganglia. In the example of Figure 6.4Ci (top row), basal ganglia input is suppressed in examples 1 and 3 and active in Examples 2 and 4. Thus, examples 2 and 4 show corticocortical communication only via the direct pathways, whereas examples 1 and 3 show such communication via both direct and transthalamic pathways. The patterns are reversed for the example in Figure 6.4Cii (bottom row). The point is that the activity patterns of basal ganglia innervation of the motor thalamus can serve to determine which cortical areas are in communication via transthalamic circuitry.

A final point can be made about basal ganglia gating of thalamus. Because, when active, this input powerfully inhibits by hyperpolarizing thalamic neurons, it follows that this deinactivates T channels in these neurons (see Chapter 3), and so when the basal ganglia input ceases and the cell repolarizes, a rebound burst is evoked (Person and Perkel, 2005; Goldberg and Fee, 2012; Kim et al., 2017). This can be seen as a "wake-up call" to the cortical target of the bursting neuron that the transthalamic circuit is now open and ready for business.

6.3.5 Some Thoughts on Terminology with Respect to Driving Inputs

Consider the following definition for a relay cell of the lateral geniculate nucleus: it is a cell that receives driving input from the retina and projects to cortex. This seems reasonable, but there are some perhaps surprising consequences of this definition.

First, there is the so-called *retinorecipient zone of the pulvinar* (Berman and Jones, 1977; Kawamura et al., 1979; Leventhal et al., 1980). But our definition of a geniculate relay cell renders this term an oxymoron because any thalamic retinorecipient zone, by our definition, is part of the lateral geniculate nucleus. Presumably, this is why Guillery and colleagues referred to this region as the "geniculate wing" (Guillery et al., 1980).

Second is the issue of geniculate layer C3 in the cat and other carnivores because this layer receives no retinal input (Hickey and Guillery, 1974; Linden et al., 1981). By our definition, a thalamic region not in receipt of retinal input cannot be considered part of the lateral geniculate nucleus. Because the C layers receive input from the superior colliculus, this means that layer C3 is innervated by the superior colliculus and not the retina. Logic thus suggests that layer C3 should be considered part of the pulvinar and not the lateral geniculate nucleus.

Some may regard this as merely semantics, but it does raise another possibility. That is, we argued above that the C layers in the cat, which is where the W cells reside, are homologous to the primate koniocellular layers and the shell region of the mouse lateral geniculate nucleus. Perhaps some koniocellular geniculate cells in the primate and shell geniculate cells in the mouse receive no retinal input but are driven instead by input from the superior colliculus.

6.4 Summary and Concluding Remarks

The recognition of the existence of higher order thalamic relays and their distinction from first order relays has proved to be an important step in further understanding of thalamic and cortical functioning. Perhaps most importantly, it provides a heretofore undiscovered route of corticocortical communication via transthalamic circuitry that must be accounted for. This transthalamic route can be gated or modulated by thalamus, offering a level of control of information flow not available to direct corticocortical connections. Furthermore, the appreciation of higher order thalamic relays, which appear to occupy most of the thalamic volume, provide a relatively straightforward role (i.e., serving as a link in transthalamic communication) for much of thalamus that had previously been rather mysterious in functional terms. Finally, the appreciation of higher order relays also puts an end to thinking of thalamus chiefly as merely a means of relaying peripheral information to cortex.

6.5 Some Outstanding Questions

1. Regarding the organization of direct and transthalamic corticocortical connections, both feedforward and feedback, we repeat the three main questions: How common is the parallel organization in which both types of pathway are involved in corticocortical communication? What is different in the nature of messages transmitted by the two pathways? Why is one of the pathways sent through the thalamus?
2. Are layer 5 inputs to thalamus always drivers there?
3. Do higher order cortical areas receive their major driver afferents from higher order thalamic relays or from other cortical areas? How do these thalamocortical and corticocortical drivers interact at their target zones? Do they converge onto the same cells, or target separate populations, or a combination of both?
4. Are the thalamocortical properties of first order and higher order relays generally the same or different?

5. Is there a good example in any species of a first order thalamic relay that does not have a functionally corresponding or related higher order relay?

6. Do feedforward axons from higher order thalamic nuclei branch to innervate multiple cortical areas, and, if so, do action potentials traveling down these branching axons evoke synchronous responses in their target areas? What purpose might this synchrony serve?

7. Are there relay cells among geniculate koniocellular cells in the monkey or W cells in the cat that relay input to cortex from the superior colliculus and not from the retina?

7

Thalamic Circuitry

An important proviso to our description of thalamic circuitry is that the best-studied such circuitry remains that of the cat's lateral geniculate nucleus. However, work on the cat's lateral geniculate nucleus has come to a virtual stop, and thus we have not much advanced our knowledge of thalamic circuitry in the past decade or so. Relevant work today centers on the use of mice or monkeys, and it will be some time before study of the thalamus in these species catches up to the knowledge base amassed for the cat. Therefore, the cat's lateral geniculate nucleus still serves as a useful general model for the mammalian thalamus and will serve that purpose here.

7.1 Inputs to Geniculate Relay Cells

Figure 7.1A summarizes the main inputs to neurons involving geniculate circuitry. Where known, it also shows the postsynaptic receptors involved, ionotropic or metabotropic, as well as the rough percentage of synapses formed by each input onto relay cells. There are several features of Figure 7.1 worth elaborating.

First, as pointed out in Chapter 5, retinal inputs activate only ionotropic receptors, whereas other inputs to this circuitry so far identified activate both ionotropic and metabotropic receptors. Second, as shown, the feedback from cortical layer 6 can directly excite relay cells and also indirectly inhibit them via activation of interneurons or reticular cells, and so it is difficult to predict the effect of activation of this feedback pathway (this is further elaborated in the description of Figure 7.1B,C in the following paragraph). This does not seem to be true for the bulk of the brainstem modulatory inputs, activation of which unambiguously excites relay cells by direct excitation and disinhibition, the latter via inhibition of the inhibitory interneuron and reticular cell inputs to relay cells. For example, the cholinergic input achieves this neat trick by activation of different muscarinic (metabotropic) receptors on each cell type: onto relay cells, M1 receptors that excite, and onto the inhibitory GABAergic cells, M2 receptors that inhibit (McCormick and Prince, 1986; McCormick and Pape, 1988; McCormick, 1989, 1992) (see Figure 4.4). Thus, when these cholinergic inputs

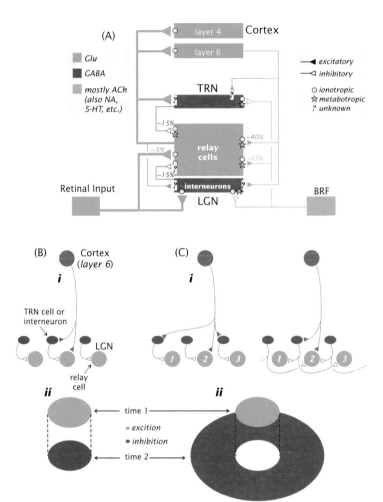

FIGURE 7.1 Details of circuitry involving the lateral geniculate nucleus. A. Schematic view of major circuit features of the lateral geniculate nucleus with related postsynaptic receptors present on relay cells. Other thalamic nuclei seem to be organized along the same general pattern, the main difference being a different driver input to replace the retinal input here. The key to the left indicates the major transmitter systems involved, and that to the right, whether the postsynaptic receptors involved are ionotropic or metabotropic or both or unknown (?) and whether the input is excitatory or inhibitory. The retinal input activates only ionotropic receptors (circles), whereas all nonretinal inputs activate metabotropic receptors (stars) and often ionotropic receptors as well.

Abbreviations: *5-HT*, serotonin; *ACh*, acetylcholine; *BRF*, brainstem reticular formation; *GABA*, γ-aminobutyric acid; *Glu*, glutamate; *LGN*, lateral geniculate nucleus; *TRN*, thalamic reticular nucleus.

B, C. Two patterns among others possible for corticothalamic projection from layer 6 to cells of the thalamic reticular nucleus, geniculate interneurons, and geniculate relay cells. B. Shown is the pattern of direct excitation and disynaptic inhibition. C. Shown is a more complicated pattern in which activation of a cortical axon can excite some relay cells directly and inhibit others through activation of reticular cells. Further details in text.

Abbreviations as in A.

are relatively silent, as happens during drowsiness, relay cells are hyperpolarized, less responsive, and more bursty, and when the cholinergic inputs are more active, as happens during vigilance, relay cells are depolarized, more responsive, and less bursty (Contreras and Steriade, 1995; Datta and Siwek, 2002; Steriade and McCarley, 1990, 2005; Bezdudnaya et al., 2006; Stoelzel et al., 2009).

Figure 7.1B,C illustrates the importance of understanding the circuitry of the corticogeniculate feedback at the single cell level. Figure 7.1B shows the traditional view of the circuitry. Here, activation of the layer 6 cell provides a simple direct excitation of the target relay cell followed by disynaptic inhibition via a thalamic reticular cell and/or interneuron. Figure 7.1C depicts subtly different circuits with the same general effect, in which activation of a particular layer 6 cell will excite one relay cell (*cell 2*) and disynaptically inhibit its neighbors (*cells 1* and *3*).

The circuit of Figure 7.1B might have little overall effect on the target relay cell's membrane voltage if the excitation and inhibition are reasonably matched, but the resultant increase in synaptic conductance will render the relay cell less responsive to other inputs (Chance et al., 2002). Thus, this circuit could act to control the gain of retinogeniculate transmission. The circuits of Figure 7.1C suggest a very different function for activation of the feedback: *cell 2* would be depolarized, rendering it more excitable and more likely to fire in tonic mode, whereas *cells 1* and *3* would be hyperpolarized, rendering them less excitable and more likely to fire in burst mode. Evidence for the circuits of Figure 7.1C does exist (Wang et al., 2006; Lam and Sherman, 2010). It is important to note that using large-scale activation or inactivation of the layer 6 feedback that reflects enough of the retinotopic map will obscure any circuitry depicted in Figure 7.1C; that is, large-scale activation of the feedback in Figure 7.1C would effectively produce monosynaptic excitation and disynaptic inhibition on each cell as in Figure 7.1B, and, likewise, large-scale inactivation would appear to remove monosynaptic activation and disynaptic inhibition from each cell.

7.2 Pattern of Synaptic Inputs to Relay Cells

Details of inputs to geniculate relay cells in the cat have been worked out largely through the use of electron microscopy, which remains the only anatomical technique to unambiguously identify synapses. Triadic circuitry in the thalamus, which was briefly described in Chapter 2, is more fully discussed here. In the cat's lateral geniculate nucleus, triads are related to X relay cells and not Y relay cells as described in Chapter 11; any involvement of triads with W relay cells remains to be determined.

7.2.1 Conventional Triadic Circuitry

A special synaptic arrangement common to thalamus but not seen in cortex is the *triad*. The classic triad involves three synapses, and an example for the cat's lateral geniculate nucleus is shown in Figure 7.2. One of the synaptic elements of a triad is quite unusual: it is a

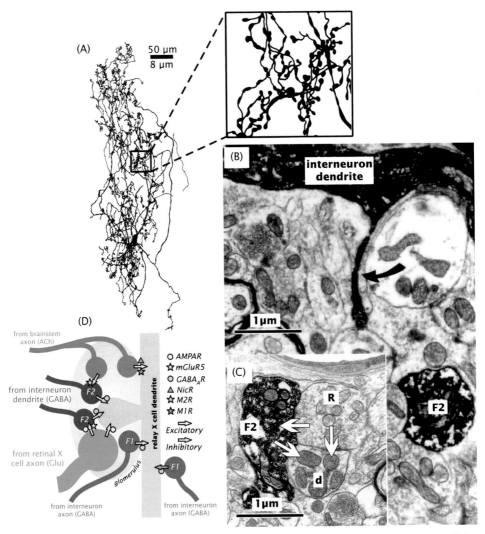

FIGURE 7.2 A–C. Interneuron from the lateral geniculate nucleus of a cat. Preparation by intracellular injection with horseradish peroxidase. Redrawn from (Hamos et al., 1985). A. Tracing of the interneuron. The inset at the top shows an enlarged view of axoniform dendritic terminals. B. Electron micrograph of a fine process (arrow) emanating from a secondary dendrite (*interneuron dendrite*) of the interneuron. This process arborizes into numerous dendritic terminals, and one is shown here (*F2*). C. The electron micrograph shows a triad in one section, with the labeled F2 terminal from the interneuron indicated. Arrows point to the three synapses of the triad: from the F2 terminal to the dendritic appendage of an X relay cell, from the retinal terminal (R) to the F2 terminal, and from the retinal terminal to the dendritic appendage. D. Synaptic inputs in and near a glomerulus. Shown are the various synaptic contacts (arrows), whether they are inhibitory or excitatory, and the related postsynaptic receptors. The conventional triad includes the lower interneuronal dendritic terminal and involves three synapses (from the retinal terminal to the dendritic terminal, from the retinal terminal to an appendage of the X cell dendrite, and from the dendritic terminal to the same appendage). Another type of triad includes the upper interneuronal dendritic terminal and also involves three synapses: a branched (cholinergic) brainstem axon produces one synaptic terminal onto an X cell relay dendrite and another onto the dendritic terminal, and a third synapse is formed from the dendritic terminal onto the same relay cell dendrite. *Abbreviations*. As in Figure 7.1 plus *ACh*, acetylcholine; *AMPAR*, α-amino-3-hydroxy-5-methyl-4-isoxazole propionic acid receptor; *d*, dendrite; *GABA$_A$R*, type A GABA receptor; *M1R & M2R*, different muscarinic receptors; *NicR*, nicotinic receptor; *R*, retinal terminal.

dendritic terminal from an interneuron. As noted in Chapter 2, an interneuron has two synaptic outputs: a traditional one from its axon and an unusual one from peripheral dendrites (Guillery, 1969b; Ralston, 1971; Famiglietti and Peters, 1972; Wilson et al., 1984; Hamos et al., 1985). The "axoniform" appearance of the dendrites is shown in Figure 7.2A, and the inset shows dendritic terminals at higher power. Figure 7.2B,C shows electron micrographs depicting the triad, and, for clarity, the interneuron had been filled with an electron-dense dye, which is why its synaptic terminal, labeled F2[1] is dark. The three synapses shown in Figure 7.2C are from the retinal terminal to a relay cell dendritic appendage, from the same retinal terminal to the interneuron (F2) dendritic terminal, and from the interneuron terminal to the same relay cell target. Thus, the dendritic terminal of the interneuron is both presynaptic and postsynaptic.

Figure 7.2D shows a schematic of this triad with added information regarding the postsynaptic receptors involved (Cox et al., 1998; Cox and Sherman, 2000; Govindaiah and Cox, 2004). The triadic arrangements shown are located within glomeruli, which include a number of other synaptic arrangements. Glomeruli are found throughout thalamus (reviewed in Sherman and Guillery, 2013), always or almost always identified with triads, but they are not found elsewhere in the central nervous system.[2] Because of their relationship to triads, glomeruli in the cat's lateral geniculate nucleus are largely limited to circuitry involving X relay cells. An unusual feature of glomeruli is that the entire structure is enveloped in a glial sheath, but the individual synapses within are not, whereas most other synapses in the central nervous system are individually wrapped in glial membranes. What this means functionally for glomeruli remains obscure.

As shown in Figure 7.2D, for the conventional triad, the GABAergic input to the relay cell activates $GABA_A$ receptors, the retinal input to the relay cell activates ionotropic glutamate receptors (mostly AMPA receptors), but the input from the retinal terminal to the interneuronal terminal activates both ionotropic and metabotropic glutamate (mGluR5) receptors. Consider how this arrangement affects retinogeniculate transmission as the retinal input increases its firing rate. Low rates will activate ionotropic but not metabotropic glutamate receptors, and this will lead to a monosynaptic EPSP followed by a disynaptic IPSP, each fairly brief. As the retinal axon increases firing above roughly 10 spikes/sec, activation of the metabotropic receptors on the inhibitory terminal ensues, with more such activation with increasing retinal firing rates (Viaene et al., 2013). This leads to prolonged further depolarization of the inhibitory terminal, which, in turn, means more GABA output. In other words, the inhibition relative to excitation seen at the relay cell increases with increased retinal firing levels. Furthermore, this increased inhibition would outlast by about 4 sec the retinal input because of the long EPSP evoked on the

1.   This is somewhat arcane terminology. The "F" reflects the idea that GABAergic terminals such as this tend to have flattened vesicles when compared to the more spherical vesicles in excitatory terminals. F1 terminals are the more standard type associated with axons, and these are to be distinguished from dendritic terminals, called F2.

2.   Structures known as glomeruli are also found elsewhere in the brain, such as in the olfactory bulb and cerebellum. But these are very different structures not to be confused with one another or with thalamic glomeruli. The common terminology is unfortunate.

inhibitory terminal by activation of the metabotropic receptor (Figure 4.4B) (Govindaiah and Cox, 2004).

We can relate this to visual responsiveness as follows (Sherman, 2004). Retinal firing rates are monotonically related to the level of contrast in the visual stimulus. As noted above, as retinal firing increases (with increased contrast), there is relatively more inhibition of the relay cell, and thus the gain of retinogeniculate transmission is reduced. This serves to increase the dynamic range of information transmission by allowing higher contrasts to be relayed without saturating relay cell responses. In other words, at high contrast, contrast gain is relatively low. Furthermore, if the scene suddenly becomes less contrasty, the gain of retinogeniculate transmission and thus contrast sensitivity would take a few seconds to be set to a higher level.

This is exactly what happens psychophysically during contrast adaptation or gain control. Contrast gain control is an important property of the visual system that, like other forms of adaptation (e.g., to brightness or motion), helps to adjust the sensitivity of visual neurons to ambient levels of stimulation. The point here is that triadic circuitry in this way provides one neuronal mechanism for this perceptual process, although it is thought that such mechanisms also exist in retina and cortex (Sclar et al., 1989; Carandini and Ferster, 1997; Sanchez-Vives et al., 2000a,b; Solomon et al., 2004).

As noted, triads like that shown in Figure 7.2 are quite ubiquitous throughout thalamus, but the details of their operation have not been much studied outside of the example described here. Furthermore, there are different relay cell types in thalamus, and triads seem to be a property of some types and not others. Indeed, even in the cat's lateral geniculate nucleus, triads seem to be a feature largely limited to X relay cells but not Y relay cells (Wilson et al., 1984; Hamos et al., 1987); on Y cells retinal terminals contact dendrites with simple, conventional synapses. A similar relationship between triadic involvement and cell type is also seen in the lateral geniculate nucleus of the rat (Lam et al., 2005).

7.2.2 Functional Pseudo-Triadic Circuitry Involving Cholinergic Brainstem Input

Figure 7.2D also shows the involvement of brainstem cholinergic inputs within glomeruli (Erisir et al., 1997a; Cox and Sherman, 2000). They seem often to form a sort of functional triad, taking the place of a retinal terminal in the arrangement, although instead of producing one synaptic terminal in the local circuit, it produces two participants: one cholinergic terminal synapses onto an F2 terminal that, in turn, contacts the relay cell, and the other cholinergic terminal contacts the same relay cell. However, since the cholinergic terminals derive from the same axon, and because the synapse onto the F2 terminal involves metabotropic receptors, the function of this circuit may closely resemble that of the conventional triad involving a retinal terminal. One important difference is that it is an M2 receptor on the F2 terminal that is involved, and the presumed action of activating this receptor would be to reduce GABA release onto the relay cell dendrite, thereby disinhibiting the relay cell (Cox and Sherman, 2000).

7.2.3 Overall Pattern of Synapses onto Relay Cells

Figure 7.3 summarizes the pattern of synaptic inputs to X and Y relay cells of the cat's lateral geniculate nucleus (Wilson et al., 1984; Erisir et al., 1997b; Van Horn et al., 2000). All the dendrites in this illustration are collapsed to one for each cell, and triads for the X cell are represented schematically. Each cell receives roughly 2,500–5,000 synapses, and the distributions shown here reflect the relative number of each input. Note the rather dramatic difference in proximal versus distal dendritic targeting: proximally are found inputs from retina, interneurons, and the brainstem reticular formation and distally are found inputs from layer 6 of cortex and the thalamic reticular formation.

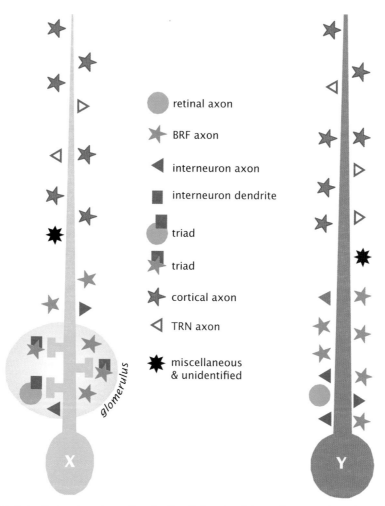

FIGURE 7.3 Synaptic inputs onto an X and a Y cell. For simplicity, only one, unbranched dendrite is shown. Synaptic types are shown in relative numbers.

Abbreviations: *BRF*, brainstem reticular formation; *TRN*, thalamic reticular nucleus. Redrawn from Sherman and Guillery, 2013.

7.2.4 Push-Pull Nature of Contribution from Interneuron Circuitry

There is an interesting relationship between the receptive fields of interneurons and the relay cells they innervate: they have opposite signs regarding their center/surround properties. Therefore, the axon of an interneuron with an on-center receptive field innervates a relay cell with an off-center receptive field, and vice-versa (Wang et al., 2011). This provides a "push-pull" nature to a relay cell's receptive field, enhancing the relationship between the on and off regions.

Note that the point of the receptive field of the interneuronal axon and its postsynaptic relay cell target having opposite signs in their center/surround organization does not apply to the dendritic output of interneurons involved in triads. We suggested above that the receptive field of the interneuron is imparted to the cell by proximal retinal inputs, and the distal dendritic outputs are so electrotonically distant from these retinal inputs as to be little affected by them. Instead, the dendritic outputs are driven by the retinal input in the triad, and so, to the extent that these dendritic outputs have a receptive field, it will be the same as that of the retinal input in the triad that also provides the receptive field of the relay cell.

Figure 7.4 explores further details of this circuitry. The figure depicts two different parallel streams of retinogeniculate processing that are labeled X and Y; we describe the significance of this more fully in Chapter 11. For the on-center relay X cell on the left, there are at least two relevant retinal axonal inputs. One (labeled 2) that participates in triadic innervation of the relay cell has an on-center receptive field, which it transfers to the relay cell. The other (labeled 1) that provides the input to an interneuron that innervates the relay cell has the opposite sign of receptive field, and the two receptive fields of retinal inputs 1 and 2 overlap in visual space. The relationship of the dendritic contribution to the triad (red terminal) to the axonal contribution is unclear: Do they come from the same interneuron or from interneurons with similar on- or off-center receptive fields? Thus, this on-center relay X cell receives axonal input from an off-center interneuron. For an off-center relay X cell (not shown), the overlapping receptive fields of the equivalent of retinal axon 1 would

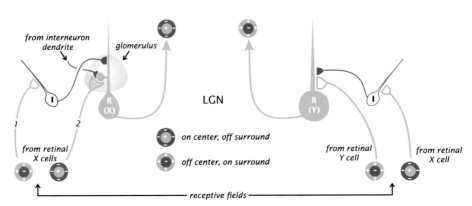

FIGURE 7.4 Figure showing "push-pull" circuitry involving relay cells and interneurons of the cat's lateral geniculate nucleus (*LGN*). Details in text.

be off-center and that of retinal axon 2 on-center. For the off-center Y cell on the right, the overlapping receptive field of the interneuron that innervates it is on-center, and vice-versa for an on-center Y cell (not shown). Note that the Y-innervating interneuron is innervated by a retinal X axon because all interneurons seem to have X-type receptive fields (Sherman and Friedlander, 1988). A proviso to this last point is evidence that Y retinal axons contact interneurons but not in triadic arrangements, which seem limited to the X pathway (Datskovskaia et al., 2001); perhaps these inputs are insufficient to provide significant Y-like properties to the receptive fields of these targeted interneurons.

Given the overlapping and opposite signs of receptive fields for the relay cells and interneuronal axonal inputs shown in Figure 7.4, the following is to be expected. Consider the X cell first. A centered spot of light will not only excite the relay cell by activating retinal axon 2, but it will also disinhibit the relay cell by inhibiting retinal axon 1. Likewise, a centered dark spot will not only reduce excitability of the relay cell by inhibiting retinal axon 2, but it will also inhibit the relay cell by exciting retinal axon 1. The same properties would play out for the Y cell shown in Figure 7.4. Thus, the circuitry of Figure 7.4 would enhance both excitation and inhibition of relay cells, and this is the property of what has been called "push-pull" circuitry in the lateral geniculate nucleus (Wang et al., 2011).

7.2.5 Variations from the Model of the Cat's Lateral Geniculate Nucleus

The lateral geniculate nucleus of the cat is a first order thalamic relay, and there have been limited studies of properties described above for other first order relays. These generally support the conclusion that the cat's lateral geniculate nucleus remains a useful model for first order relays (reviewed in Jones, 2007; Sherman and Guillery, 2013). However, one example of the limitation of this model is that the description of interneurons just given reflects only those found in the A layers of the cat's lateral geniculate nucleus, and it is now clear that other classes of interneuron exist (Carden and Bickford, 2002). Clearly, we need more examples of thalamic circuitry and cell types to appreciate the full range of relay properties that exist.

7.3 Differences Between First and Higher Order Thalamic Relays

To a first approximation, the main difference between first and higher order thalamic relays is the origin of the driver input: subcortical or layer 5 of cortex (Figure 6.1). Non-driver circuitry is fairly similar in these two relay classes. However, a deeper inspection of circuitry reveals a number of subtle but likely important differences.

7.3.1 Relative Number of Driver Synapses

An analysis of first and higher order sensory thalamic nuclei in the cat, including the visual, somatosensory, and auditory relays, has revealed a dramatic difference between the two types of relay. That is, higher order nuclei have a much lower percentage of driver synapses than do first order nuclei, approximately 2% versus approximately 7% (Van Horn et al., 2000; Wang et al., 2002; Van Horn and Sherman, 2007). These data represent relative numbers of driver synapses, so whether these differences reflect fewer overall driver synapses in higher order nuclei, more non-driver synapses, or both, remains to be determined.

7.3.2 Origin of Non-Driver Inputs

Some sources of non-driver input appear to target higher order and not first order nuclei. In the rat, both the zona incerta and anterior pretectal nucleus target certain higher order nuclei fairly exclusively with GABAergic input (Power et al., 1999; Barthó et al., 2002; Lavallée et al., 2005). Furthermore, the GABAergic input from the globus pallidus (Sakai et al., 1996; Kuramoto et al., 2011) and substantia nigra (Carpenter et al., 1976; Tanibuchi et al., 2009) appears to selectively target higher order relays. Finally, in monkeys and humans, dopaminergic input from various subcortical sources selectively target higher order thalamic nuclei (Sanchez-Gonzalez et al., 2005; Garcia-Cabezas et al., 2007).

Taken together, these data suggest that higher order thalamic relays are under additional sources of modulatory control as compared to first order relays, and this might relate to the smaller ratio of driver inputs seen in these higher order nuclei. In particular, higher order thalamic nuclei receive substantial GABAergic inputs from the zona incerta, substantia nigra, basal ganglia, and pretectal region, and these GABAergic pathways do not extensively innervate first order nuclei (Sakai et al., 1996; Bokor et al., 2005; Lavallée et al., 2005; Gulcebi et al., 2011; Kuramoto et al., 2011). This implies that gating of higher order relays can be selectively and more effectively controlled than is the case for first order relays, which would affect transthalamic corticocortical communication.

As noted in Chapter 6, for the GABAergic basal ganglia input to thalamus, these extra GABAergic inputs to higher order thalamic relays, when active, can interrupt targeted transthalamic pathways. Furthermore, as also suggested for the basal ganglia input, the inhibition of relay cells, if it lasts long enough, will de-inactivate the T-type Ca^{2+} channels in the relay cells. Thus, when the inhibition ceases and the transthalamic gateway opens, the first significant input from layer 5 to the affected thalamic neurons will elicit a burst, with the potential significance of such a response, such as providing a wake-up call to the targeted cortical area as suggested later in Chapter 10.

7.3.3 Response to Classical Modulatory Inputs

In recordings from the rat, first and higher order sensory thalamic nuclei—respectively, the lateral geniculate nucleus and pulvinar for vision, the ventral posterior and posterior medial nuclei for somatosensation, and the ventral and dorsal regions of the medial geniculate nucleus for audition—differences were found in responses of neurons to modulatory input. Specifically, serotonergic and cholinergic inputs from the brainstem depolarize all first order relay cells but a significant minority (one-fourth to one-third) of those in higher order nuclei are hyperpolarized by these inputs due to different receptors to these neurotransmitters (Varela and Sherman, 2007, 2008).

This means that, whereas these modulatory systems would depolarize all first order relay cells and thereby increase responsiveness and bias them toward tonic firing, these modulators would hyperpolarize a substantial fraction of higher order relay cells, thereby decrease responsiveness and biasing them toward burst firing. This bias toward burst firing would also be expected from the additional GABAergic inputs to higher order thalamic relay nuclei, as described above. From these observations, one would predict that burst firing is more prevalent in higher order nuclei compared to first order, and evidence for this has been presented in monkeys (Ramcharan et al., 2005).

7.3.4 Possible Convergence of Driver Input to Relay Cells

First order relay cells generally transfer information from one or a few driver inputs without further significant elaboration of the information carried (Cleland et al., 1971; Usrey et al., 1999; Rathbun et al., 2010b); but see (Bickford et al., 2015; Morgan et al., 2016; Litvina and Chen, 2017). However, limited evidence exists for such elaboration for some higher order relay cells, where single neurons in the posterior medial nucleus or pulvinar are innervated by both layer 5 and subcortical driver inputs (Groh et al., 2014). This is a critical issue and needs indisputable confirmation, because current ideas of thalamic processing do not include elaboration of information based on significant convergence of driver inputs. However, a recent study has demonstrated significant convergence in the mouse of layer 5 (driver) inputs from the primary somatosensory and motor cortices to individual thalamic neurons of the posterior medial nucleus (Sampathkumar et al., 2021).

7.3.5 Relationship to Thalamic Reticular Nucleus

Recent evidence points to variability in cell and circuit properties within the thalamic reticular nucleus (Clemente-Perez et al., 2017; Li et al., 2020; Martinez-Garcia et al., 2020). Of particular interest, this includes different circuits for first and higher order thalamic relays. For instance, regarding the somatosensory thalamus in mice, Martinez-Garcia et al. (2020) have shown that calbindin-containing reticular cells, which occupy central core regions of the thalamic reticular nucleus, innervate the first order ventral posterior nucleus, whereas somatostatin-containing reticular cells, which occupy the surrounding edges of the nucleus, innervate the higher order posterior medial nucleus.

7.4 Higher Order "Nuclei" Versus Higher Order "Relays"

First order thalamic nuclei seem completely first order in the sense that all driving input to their relay cells emanates from subcortical sites. In contrast, higher order nuclei appear often to contain first order elements. Examples are the pulvinar, mediodorsal nucleus, and posterior medial nucleus, all of which are generally considered to be higher order nuclei. The pulvinar and medial dorsal nucleus are both innervated by the superior colliculus (Beltramo and Scanziani, 2019), a subcortical source, and if this collicular input contains driver afferents (Kelly et al., 2003), then these higher order nuclei would also contain first order circuitry; the posterior medial nucleus has some driver input from the spinal nucleus of the fifth nerve (Mo et al., 2017). The degree of mixed first and higher order circuits in some thalamic nuclei can lead to terminological confusion. In this regard, it may be better to refer to order in terms of the basic functional unit—namely, the relay cell—rather than the nucleus and thereby refer to first and higher order relays rather than to nuclei. However, given the possibility of convergence of driver inputs to some relay cells discussed above, some relay cells may receive convergent driver input from a subcortical source and layer 5 of cortex. The distinction between a first order relay and a higher order one remains important, but the issues of how they play out at the single relay cell level need be further explored and kept in mind.

7.5 Summary and Concluding Remarks

A key feature of thalamic circuitry is the lack of connectivity between relay cells. The only local inputs to relay cells derive from GABAergic interneurons and reticular cells, and, as noted in Chapter 2 and Figure 7.1, these contribute only about 30% of the synapses onto relay cells (Van Horn et al., 2000). This means that more than two-thirds of the synaptic input to relay cells arrives from external sources, which is very different from cortical circuitry, in which the vast majority of synaptic inputs to neurons are local (see Chapter 8). It is thus the case that the relay through thalamus involves minimal further processing beyond local inhibition and the modulatory effects of external inputs.

There does seem to be a fairly common plan for thalamic circuitry across nuclei and mammalian species, and this is best exemplified by that of the cat's lateral geniculate nucleus (Figure 7.1A). However, differences do exist among nuclei, and, for the most part, this is an unexplored issue that needs more data. There also appears to be important differences between the details of circuitry in first order versus higher order thalamic relays. In general, these differences amount to more modulatory circuitry for the latter and especially more extrathalamic inhibitory control, factors that likely influence transthalamic corticocortical communication. In particular, the extra inhibitory input could gate these transthalamic circuits and thereby determine which cortical areas are in communication only by their direct connections versus those communicating via both direct and transthalamic circuits.

7.6 Some Outstanding Questions

1. What is the function of the triad, and why do they seem limited to just some relay cell classes?
2. What is the functional significance of thalamic glomeruli within which individual synapses are not enwrapped by glia?
3. Given that some relay cells are innervated both by axons and dendritic terminals of interneurons (e.g., X cells), what is the relationship of the interneurons involved, and are they ever or always the same interneurons?
4. What is the significance of the difference in circuitry between first and higher order thalamic nuclei, especially the extra modulatory and inhibitory inputs to the latter?
5. Are there any pure higher order thalamic nuclei? Or are all higher order relays mingled with first order pathways?
6. How functionally isolated are dendritic F2 outputs of interneurons from each other and from the soma? Are there any interneurons that lack action potentials? Do the action potentials of interneurons invade the dendrites, affecting the F2 terminals? If they do, what is the functional effect?

8

Brief Overview of Cortical Organization

A major hallmark of the mammalian brain is the cerebral cortex[1] and, with it, thalamocortical interactions. With cortical evolution, sensory, motor, and cognitive activities that were once the sole domain of the brainstem were expanded upon with new circuitry that allowed for a greater range of neural computations and behaviors. Across mammals, the size of the cerebral cortex and the number of cortical areas vary tremendously; however, many cortical features, including cell types, lamination, input/output organization, and intrinsic circuitry, are largely conserved. In the following sections, we provide a brief overview of cortical organization emphasizing properties that are conserved across species and cortical regions as well as noteworthy differences. When relevant, details about cortical cell types are provided; however, an expanded description of cell types in the cortex is provided in Chapter 2.

8.1 Cortical Expansion Built on a Common Plan

As a general rule, brain size increases with body size. However, a doubling in body weight does not equate to a doubling in brain weight. Rather, small animals typically have a higher brain-to-body weight ratio compared with larger animals. It is therefore noteworthy that the brain-to-body weight ratio of a mouse is similar to that of a human, thus indicating an expansion in the brain weight of humans compared to other similarly sized mammals. To compare extremes, at less than 2 g in total weight, the Etruscan shrew is the smallest terrestrial mammal and it likely has the smallest cerebral cortex. Still, the cortex of the Etruscan shrew contains multiple cortical regions and areas that have similar cytoarchitectonic and histological features as seen in larger mammals (Naumann, 2015), including the African elephant

---

1. Although the cerebral cortex includes archicortex (hippocampus) and paleocortex (pyriform or olfactory cortex), it is the neocortex that is highly interconnected with the thalamus and thus the focus of this chapter. Unlike neocortex that has six characteristic layers, archicortex and paleocortex have just a few layers.

that weighs up to 7.5 tons and has a brain that can weigh up to 6 kg (Haug, 1987; note that the human brains weighs ~1.5 kg).

8.2 Cortical Areas

There was a bit of a cottage industry among neuroanatomists of the 19th and early 20th centuries to define different cortical areas in the human brain based largely on histological differences, and the number of distinct areas claimed ranged from five to several hundred. One such effort in particular, that of the German anatomist Korbinian Brodmann (1868–1918), emerged as the dominant map of these cortical areas (Brodmann, 1909; Figure 8.1). In this map, cortical areas were assigned numbers (e.g., area 17 for primary visual cortex; area 4 for primary motor cortex; and areas 3, 1, and 2 for primary somatosensory cortex), and the same nomenclature was used for matching areas between species. Remarkably, much of Brodmann's nomenclature continues to be in use today despite general agreement that the cortex, particularly in primates, contains more than the 52 areas identified by Brodmann. Although Brodmann is best known for distinguishing cortical areas, he also established the now common view that the cerebral cortex is comprised of six layers (described below).

Thalamic inputs also play a key role in defining cortical areas. For instance, primary sensory areas receive strong, driving feedforward input from thalamic relay cells that are themselves innervated by neurons conveying signals originating from sensory receptors. Thus, primary visual cortex, auditory cortex, and somatosensory cortex receive synaptic input from the lateral geniculate nucleus, the medial geniculate nucleus, and the ventral posterior nucleus, respectively. Moreover, within these cortical areas, thalamic inputs establish a functional architecture upon which cortical computations are performed (Figure 8.2). For instance, within primary visual cortex, a highly structured organization of synaptic input from the lateral geniculate nucleus establishes a cortical map of visual space, alternating ocular dominance columns, and columns of neurons selective for stimulus

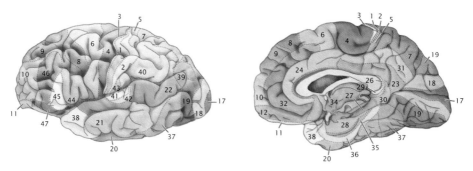

FIGURE 8.1 Illustration from Korbinian Brodmann's 1909 publication *Vergleichende Lokalisationslehre der Grosshirnrinde* (Brodmann, 1909) showing the locations of various regions in the human cerebral cortex that he defined on the basis of cytoarchitecture. *Left*, lateral view; *right*, medial view. Brodmann assigned numbers to identify 52 regions in the human brain, and many of these regions are still referred to using Brodmann's nomenclature, such as areas 17 and 18 for V1 and V2, areas 3, 1, and 2 for primary somatosensory cortex, area 4 for motor cortex, and areas 41 and 42 for parts of primary auditory cortex.

Visual Cortex

(A)

(B)

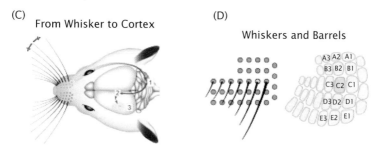

Somatosensory Cortex

(C)

From Whisker to Cortex

(D)

Whiskers and Barrels

A3 A2	A1
B3 B2	B1
C3 C2	C1
D3 D2	D1
E3 E2	E1

Auditory Cortex

(E)

(F)
corresponds to apex of cochlea   Corresponds to base of cochlea

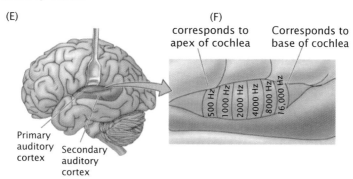

Primary auditory cortex   Secondary auditory cortex

FIGURE 8.2 Functional organization of primary visual cortex, somatosensory cortex, and auditory cortex established by thalamic inputs. A and B. Afferent axons from the lateral geniculate nucleus establish a topographic organization for retinotopy, ocular dominance, and orientation preference in primary visual cortex. A. Following visual stimulation with a bulls-eye pattern (left), monkeys given 2-deoxy-glucose show selective uptake in neurons with receptive fields distributed in a retinotopic organization across primary visual cortex (right). B. Surface view of primary visual cortex with contours indicating alternating ocular dominance columns (black lines) and the organization of orientation preference (colored lines, each hue represents preference for a particular orientation). Data acquired using intrinsic signal imaging, in vivo. C and D. Afferent axons from the ventral posterior nucleus establish a topographic organization of the body surface in primary somatosensory cortex. C. In rodents, thalamic axons carrying information from individual whiskers provide the substrate for the barrel field in cortex. D. The topography of the barrel field corresponds to the topography of the whiskers on the face. E and F. Afferent axons from the medial geniculate nucleus establish a tonotopic organization for frequency preference in primary auditory cortex. E. Lateral view of the human brain indicating the location of primary auditory cortex. F. Tonotopic organization for sound with reference to the organization of frequency preference in the cochlea. Illustrations adapted from Tootell et al., 1982; Hübener et al., 1997; Petersen, 2007; Purves et al., 2012.

orientation (Hubel and Wiesel, 1977). Likewise, structured inputs from the medial ge-niculate nucleus and ventral posterior nucleus establish tonotopic and somatotopic maps in the primary auditory and somatosensory cortices, respectively (Miller et al., 2001a,b; Winer et al., 2005; Petersen, 2007; Read et al., 2011). As a general rule, each of these areas has an overexpansion of territory, or cortical magnification, that corresponds to the re-gion of the sensory surface with maximum acuity, and this magnification is established, in large part, by the density of peripheral receptors and the resulting distribution of thal-amocortical axons. Particularly noteworthy examples of cortical magnification include the cortical representations of the fovea, hands, and face in primates; the whiskers of rodents; the frequencies used for echolocation in bats; the bill of the platypus; and the star of the star-nosed mole (Krubitzer, 2007; Petersen, 2007; Catania, 2011; Harding-Forrester and Feldman, 2018; Suga, 2018).

Higher cortical areas also have specific relationships with the thalamus that, in many cases, support a functional architecture, such as the cytochrome-oxidase rich stripes in area 18 (V2) that receive structured input from the pulvinar nucleus (Levitt et al., 1995). Whether or not all higher cortical areas have a functional architecture that reflects their thalamic input is an important and unresolved question, as is the nature of what is represented in that ar-chitecture. Still, where examined, every cortical area receives input from the thalamus and projects to the thalamus.

8.3 Cortical Layers

As mentioned above, Brodmann played a major role in establishing the notion of a cor-tical organization based on six layers. Although this six-layer "rule" sometimes requires a bit of creativity to maintain, such as in monkey primary visual cortex where layer 4 includes sublayers 4A, 4B, 4Cα, and 4Cβ, it provides a useful framework for examining the input/output organization of cortex as well as the organization of intrinsic circuits.

8.3.1 Laminar Organization of Inputs and Outputs

Each of the six cortical layers has a specific organization of inputs and outputs (Figure 8.3). In general, the feedforward input from first order thalamic nuclei to a cortical area is directed primarily to layer 4 and, to a lesser extent, the more superficial and deep layers (reviewed in Felleman and Van Essen, 1991; but see Coogan and Burkhalter, 1990, 1993), whereas that from higher order thalamus more heavily targets layer 5 (reviewed in Sherman and Guillery, 2013), and matrix thalamus (see Chapter 9) projects mainly to the superficial layers of cortex, including layer 1 (reviewed in Jones, 2007). Although layer 4 in some cortical areas and spe-cies contains projection neurons, this layer is generally an input layer. In contrast, neurons in layers 2, 3, 5, and 6 are the source of projections to stereotypical targets: neurons in layers 2 and 3 send axons to higher cortical areas, neurons in layer 5 send axons to higher order thalamic nuclei and to various brainstem and spinal cord targets, and neurons in layer 6 send

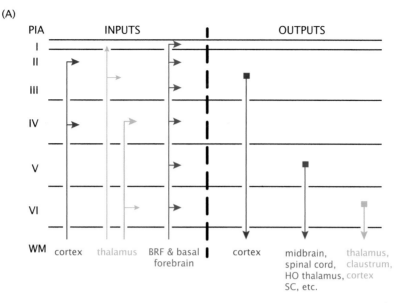

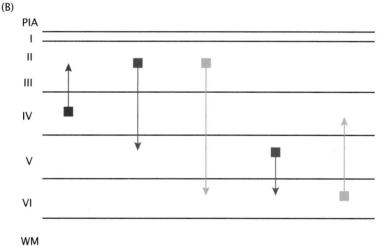

FIGURE 8.3 Circuit diagrams illustrating the general organization of cortical inputs, outputs, and interlaminar excitatory connections. A. Across cortical areas, glutamatergic afferent axons from lower cortical areas terminate primarily in layer 2, whereas glutamatergic axons from the thalamus terminate in primarily in layers 4 and 6 as well as to layers 1–3. Cholinergic, noradrenergic, etc., axons from the basal forebrain terminate in all cortical layers. Layer 3 pyramidal cells send glutamatergic axons to higher cortical areas, whereas layer 5 glutamatergic cells send axons to subcortical targets including the midbrain, spinal cord, superior colliculus, and higher order thalamus. Different classes of pyramidal neurons in layer 6 send axons either to the thalamus or to the claustrum. Note: the only pathway for cortical information to leave the cortex is via the axons of layer 5 or 6 cells. B. General organization of intrinsic excitatory connections in cortex. Across cortical areas, there is a general plan where excitatory neurons in layer 4 provide local input to layers 2 and 3, layers 2 and 3 neurons provide input to layers 5 and 6, layer 5 neurons provide input to layer 6, and layer 6 neurons provide input to layer 4.

axons to all thalamic nuclei (first order and higher order) and to the claustrum (Kelly and Gilbert, 1975). Exceptions to this general scheme exist, particularly in rodents (D'Souza and Burkhalter, 2017); nevertheless, in all cases, projection neurons are glutamatergic and, consequently, generally excite downstream target neurons. However, because some modulatory glutamatergic inputs can activate Group II metabotropic glutamate receptors that produce prolonged IPSPs, which have been described for the layer 6 to layer 4 input (Lee and Sherman, 2009a), glutamatergic input can sometimes inhibit. Layer 1 is mostly neuropil with a sparse distribution of GABAergic neurons. Within the neuropil are axons running parallel to the surface, some from thalamus and others from other cortical areas, and apical dendrites from the underlying cortical layers, mostly from pyramidal neurons in layers 3 and 5; also running in layer 1 are axons from outside sources that supply the area with neuromodulatory neurotransmitters, such as acetylcholine and noradrenaline (Lam and Sherman, 2019).

The stereotypical organization of feedforward and feedback projections just described has been applied broadly to the visual system in primates to establish a hierarchical model of cortical processing across the more than 30 visual cortical areas (Figure 8.4; see Felleman and Van Essen, 1991). In this model, primary visual cortex (V1) sits at the base of the cortical network, and higher cortical areas are located at progressively higher levels in the diagram. This model includes roughly five hierarchical circuits organized in parallel. With the exception of thalamic input to V1, this model and the hierarchy implied from it does not take into account thalamocortical circuits involving the pulvinar, which likely provide alternate routes for corticocortical communication (see Chapter 6 for more on this topic).

8.3.2 Laminar Organization of Intrinsic Cortical Circuits

An intrinsic network of excitatory cortical circuits follows a stereotypical columnar plan that is generally shared across cortical areas, with neurons in layer 4 projecting to layers 2/3 and 6, layers 2/3 neurons projecting to layer 5, layer 5 neurons projecting to layer 6, and layer 6 neurons projecting back to layer 4 (Figure 8.3B). This organizational scheme, as outlined here, is quite general. It should therefore be emphasized that species- and area-specific exceptions and elaborations on this scheme are common, as is seen in primary visual cortex of the tree shrew (Figure 8.5; Fitzpatrick, 1996). Nevertheless, the ubiquitous nature of the cortical scaffold for extrinsic and intrinsic circuits has facilitated computational models of cortical circuitry based on "conical circuits" (Figure 8.6). One aspect generally missing from these schemes is the identity of any of these connections as driver or modulator. For instance, it has been shown that the layer 4 to layers 2/3 input is driver (DePasquale and Sherman, 2012) and that from layer 6 to layer 4 and within layers 2/3 are modulator (Lee and Sherman, 2009a,b; DePasquale and Sherman, 2012), but other parts of this circuitry have so far not been identified as such.

8.3.3 Thalamic Versus Cortical Layering

There is an interesting difference in lamination that occurs occasionally in thalamus, such as in the lateral geniculate nucleus, and that seen in cortex. The six layers of cortex are a ubiquitous feature seen throughout cortex in all mammals. Geniculate layering, in contrast, can be quite variable. Three examples serve to make this point about geniculate layering. First, in closely related carnivore species, we see that on- and off-center cells are separated into

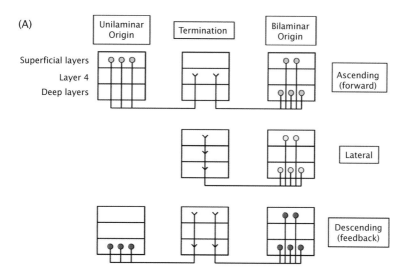

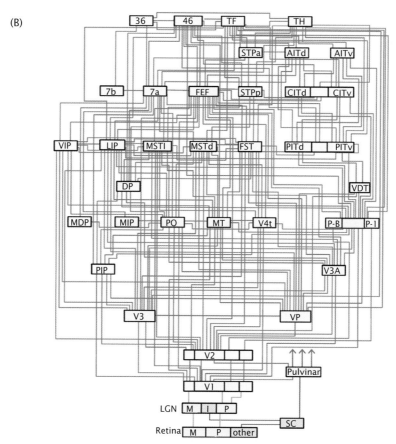

FIGURE 8.4 Schematic diagrams illustrating interareal hierarchy in the primate visual system based on anatomical projection patterns; from (Felleman and Van Essen, 1991). A. Connectivity rules to be applied to published results for assigning hierarchical relationships between cortical areas. Using these rules, feed-forward projections arise from neurons in the superficial (and sometimes superficial and deep) layers and terminate in layer 4; feedback projections arise from neurons in the deep layers (and sometimes deep and superficial) and terminate in layers outside of layer 4; and lateral projections arise from neurons in both the superficial and deep layers and terminate across the depth of the cortex. B. Circuit diagram of the visual system based on the connectivity rules shown in A. In this diagram, lower areas in the cortical hierarchy (e.g., V1) are shown lower than areas considered higher (e.g., MT) in the cortical hierarchy. Areas at similar heights are considered to be at similar a similar hierarchical level. Lines indicate anatomical connectivity.

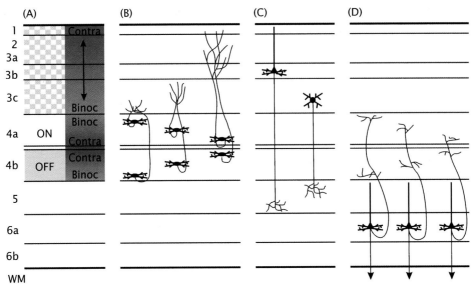

FIGURE 8.5 The intracortical circuitry in primary visual cortex of the tree shrew is an example of species- and area-specific elaboration in the precision of excitatory connections between cortical layers. A. Inputs from the lateral geniculate nucleus are organized with respect to the eye (ipsilateral vs. contralateral) and sign (on vs. off) of information supplied. Geniculate axons carrying information originating from on-center retinal ganglion cells terminate in layer 4a, whereas axons carrying information from off-center retinal ganglion cells terminate in layer 4b. Within these two division, axons carrying information originating from the contralateral eye terminate across the depth of their respective layer, whereas axons carrying information originating from the ipsilateral eye terminate near the outer edge of their respective layer. As a consequence of this organization of connections, ocular dominance has a laminar rather than columnar organization (columnar being characteristic of carnivores and primates), with cortical neurons near the outer edges of 4a and 4b having binocular responses, and neurons near the cell sparse cleft between layers 4a and 4b having contra-only responses. B. The projections from layer 4 to layers 2 and 3 serve to bring together the on and off channels, while maintaining ocular dominance. As a result, neurons in the superficial layers are cells with on/off receptive fields and ocular dominance shifts progressively from contra-only near layer 1 and binocular near layer 4. C. Likewise, the projections from layers 2 and 3 maintain the distinction in ocular dominance across layer 5. D. Inputs from layer 6 to layer 4 are selective for sublaminar regions of layer 4 with matching ocular dominance. Adapted from Fitzpatrick, 1996.

separate A layers in the mink and ferret (LeVay and McConnell, 1982; Stryker and Zahs, 1983), but these commingle in the A layers of the cat (Hubel and Wiesel, 1961). Second, different monkey species have different numbers of geniculate layers: generally six for macaque monkeys and four for owl monkeys (Le Gros Clark, 1941a,b; Kaas et al., 1978), and, in humans, the number of layers vary within an individual, as some of the parvocellular layers split into two over different parts of the nucleus (Hickey and Guillery, 1979; Jones, 2007). Third, the apparently homologous cat W, X, Y and monkey K, P, M cells have different laminar relationships: the X and Y cells commingle in the A layers, and the W cells lie separately in the C layers, whereas the M and P cells have separate layers and the K cells partially overlap M and P cells at the bottom of their respective layers (Dreher et al., 1976; Sherman et al., 1976; Wilson et al., 1976; Schiller and Malpeli, 1978; Irvin et al., 1986).

We suggest the following explanation for this. Cortical cells receive the vast majority—probably more than 95%—of their synaptic inputs from very local sources (Levy and Reyes,

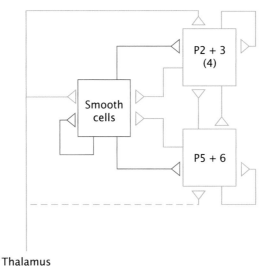

Thalamus

FIGURE 8.6 Block diagram of excitatory and inhibitory connections that successfully models the intracellular responses of cortical neurons to stimulation of thalamic afferents. Three groups of neurons interact with one another. The first group is inhibitory (smooth cells, red box and terminals), the second excitatory pyramidal cells (P) in layers 2 and 3 (green box and terminals; note this group also includes layer 4 spiny stellate cells), and the third group is excitatory pyramidal cells (P) in layers 5 and 6. Some neurons in each of the groups receive direct excitatory input from the thalamus. The thalamic input indicated with the continuous line represents stronger drive than the thalamic input indicated with the dashed line. The thickness of the lines emanating from the inhibitory smooth cells indicates strength of the inhibitory input to target cells. Adapted from Douglas and Martin, 1991.

2012; Jiang et al., 2015; Miller, 2016). We thus argue that there would have existed strong evolutionary pressure to provide efficient organization of such a locally connected structure, a plan that, once evolved, would be preserved to support the canonical organization of cortical connectivity discussed above: thus, the evolution of a relatively uniform six layers across cortex. Thalamus is quite different because only a minority—perhaps 25%—of inputs to relay cells derive from local sources and these are GABAergic (Erisir et al., 1997b; Van Horn et al., 2000). Because of a relative lack of local connectivity, evolution would have exerted less pressure to adopt a single layering plan for the lateral geniculate nucleus.

8.4 Summary and Concluding Remarks

The cerebral cortex is organized into distinct areas that have specific relationships with each other and with the thalamus. Within each cortical area, a general pattern of six layers provides a scaffold for a stereotypical organization of inputs, outputs, and intrinsic circuits. Although species-specific exceptions and elaborations exist, in general (1) feedforward inputs are directed to layer 4 and feedback projections originate from layer 6; (2) layers 2 and 3 neurons project to higher cortical areas, and layers 5 and 6 project to subcortical targets; and (3) within a cortical area, there are excitatory projections from neurons in layer 4 to layers 2, 3, and 6, from layers 2 and 3 to layer 5, from layer 5 to layer 6, and from layer 6 back to layer 4 (Lund, 1973, 1988; Gilbert and Kelly, 1975; Lund et al., 1975; Rockland and Pandya, 1979;

Felleman and Van Essen, 1991; Fitzpatrick, 1996). This canonical circuit is repeated across the cortex, indicating that it performs similar computations that manifest differently due to differences in the nature of the signals that are supplied to the cortical area.

8.5 Some Outstanding Questions

1. Does every cortical area get input from (multiple) thalamic nuclei?
2. Although assumed to be the case, does every cortical area have layer 5 and layer 6 projections to the thalamus?
3. When thalamic axons branch to innervate more than one cortical area, are the synaptic properties of the axonal inputs in cortex similar?
4. Cortical neurons receive input from many sources and provide output to many targets. In general, is the amount of convergence and divergence onto and from a cortical neuron similar?
5. How do new cortical areas evolve? Are they added de novo to the cortical sheet, do they emerge from a module that was once within a cortical area, or is a different mechanism(s) involved?

9

Classification of Thalamocortical and Corticothalamic Motifs

As noted in the beginning paragraphs of Chapter 5, a first step in analyzing complex systems, like retinal organization or circuit components, is a classification of component elements. This applies as well to thalamocortical and corticothalamic projections. That is, there is great variety in the types of thalamocortical and corticothalamic projections, and a prerequisite to any deep understanding of these critical parts of brain circuitry is developing a proper classification of their elements. How many distinct classes of thalamocortical and corticothalamic projection exist? Can the main classes be broken down into subclasses? Can we establish homologous relationships between apparently similar classes among species? What sort of functional correlates can we establish for each class? How can disease or injury related to the thalamus be associated with specific thalamocortical or corticothalamic types? This chapter includes and builds on previous discussions of such classification (Halassa and Sherman, 2019; Usrey and Sherman, 2019).

9.1 Thalamocortical Motifs

There are at least two more specific reasons to strive for a proper classification of thalamocortical motifs. First, virtually all information that arrives in cortex and thus reaches conscious perception and contributes to cognitive processing passes through thalamus. We no longer take seriously the old idea that thalamus acts as a homogeneous, machine-like relay. We appreciate that the complex cell and circuit properties of thalamus indicate that information relayed to cortex reflects considerable modulation via the thalamic relay. Any understanding of the range of such thalamocortical effects will require a full classification of thalamocortical motifs. By this we mean a complete understanding of the variety of and variability within different types of thalamocortical projection at the single-neuron level. Second, if we are to achieve a deep understanding of the broad principles of thalamocortical processing in mammals, which is a prerequisite of applying the lessons of animal research to the human condition, we will need to understand which of these organizing principles are common to all mammals and which are unique to one or a limited group of species, such as rodents. This means establishing homologies for these principles between species. Before we can assign homologies to thalamocortical motifs, we need to know how many classes of each exist for any species.

This issue of thalamocortical classification has recently been emphasized and reviewed (Halassa and Sherman, 2019), and here we summarize and expand on this assessment. Many parameters can contribute to such a classification scheme, and we break these down into three major groups: (1) based on a division of thalamus into discrete nuclei via histological criteria; (2) based on different types of driver input to thalamic relay cells; and (3) based on the nature of thalamocortical projections. Note that these sets of criteria can be combined in various ways.

There have been numerous attempts to classify thalamocortical projections using these sorts of organizational properties. Examples are core versus matrix, first versus higher order, parallel X and Y or parvocellular and magnocellular streams, etc. Figure 9.1 schematically shows limited examples of the various schemes proposed. Below, we consider these and explain why none is adequate as a complete classification scheme.

9.1.1 Classification Based on Nuclear Divisions

The first attempts to classify thalamocortical types used the distinction based on histological criteria that thalamus could be divided into separate nuclei. This classification scheme was further justified with the observation that many such classes could be defined as homologous in different species. Examples are the main sensory and motor relays (i.e., the lateral geniculate, the medial geniculate, and ventral posterior nuclei for sensory relays and the ventral anterior and ventral lateral nuclei for motor relays). Many other thalamic nuclei also appear to be homologous across species; some examples are the medial and lateral dorsal nuclei and various midline nuclei.

However, a consideration of the pulvinar shows some of the drawbacks to this approach. There are at least three problems associated with the pulvinar regarding the establishment of homologs among thalamic nuclei across species. This goes to the heart of the limitations in using thalamic nuclei as a basis for classification of thalamocortical motifs.

First, a nucleus called the pulvinar is identified in most mammals, but neither mice nor rats are said to have a pulvinar. Given that the pulvinar is the largest thalamic relay for vision, its apparent absence in mice and rats has raised doubts about the usefulness of these species as models to study vision. Mice and rats do have a nucleus named "lateral posterior" (also found in other mammals), and it now seems clear that this nucleus is indeed homologous to the pulvinar of other species (Zhou et al., 2017).

Second, Figure 1.2 does not include among its relay nuclei the posterior medial nucleus, which, as emphasized in Chapter 6, is a nucleus quite important in somatosensory processing. This nucleus has been identified in most mammals, but not in monkeys or humans, and Figure 1.2 depicts the human thalamus. We suggest that this is a terminological

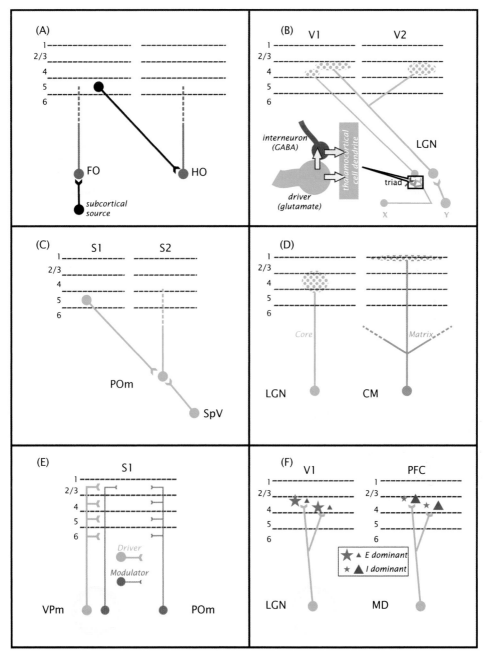

FIGURE 9.1 Limited examples of various thalamocortical motifs. A. First and higher order relays (*FO* and *HO*, respectively). First order relays receive their driving input from a subcortical source (e.g., retinal input to the lateral geniculate nucleus), whereas higher order relays receive their driving input from layer 5 of cortex. B. X and Y streams through the cat's lateral geniculate nucleus (*LGN*). On the input side, retinal Y axons form simple synapses onto dendritic shafts, whereas retinal X axons form triadic synaptic complexes. On the output side, X axons innervate only V1, chiefly in the lower half of layer 4, whereas Y axons innervate the upper half of layer 4 and branch to innervate V2 as well. C. Example of relay cell in the posterior medial nucleus (*POm*) receiving convergent driver input from both layer 5 of the primary somatosensory cortex (*S1*) as well as from the spinal nucleus of the fifth nerve (*SpV*). This

deficiency. That is, we believe that the anterior portion of the pulvinar, which has many connections with somatosensory cortex (Pons and Kaas, 1985; Cusick and Gould, III, 1990; Krubitzer and Kaas, 1992; Padberg et al., 2009), as opposed to the rest of pulvinar, which connects with visual cortex, is actually homologous with the posterior medial nucleus so identified in other mammals. Naming of this thalamic region should be changed accordingly.

Third, the pulvinar is a complex structure that can be further divided, for instance, into tectorecipient and corticorecipient zones (Zhou et al., 2017), suggesting that it really should be classified as several separate nuclei. This is true of other nuclei as well: the medial geniculate nucleus has ventral, medial, and dorsal divisions; the ventral anterior and lateral nuclei have separate cerebellar and basal ganglia recipient zones (Sakai et al., 1996; Kuramoto et al., 2013); etc. The medial geniculate nucleus is especially interesting in this regard because it contains both first and higher order divisions—namely, the ventral and dorsal divisions, respectively—whereas the somatosensory equivalents, the ventral posterior (first order) and posterior medial (higher order) nuclei are described as separate nuclei. Finally, the pulvinar includes other classification results described more fully below, such as first and higher order relays and core and matrix projections. Overall, the classification of functional thalamocortical motifs based on defined nuclear boundaries seems flawed and limited.

9.1.2 Classification Based on Driver Input

9.1.2.1 First and Higher Order Relays

First and higher order relays are identified on the basis of the derivation of their driver inputs (Figure 9.1A). As detailed in Chapter 6, first order relays receive driver inputs from subcortical sources, such as the lateral geniculate nucleus receiving driver inputs from the retina; higher order relays receive their driver inputs instead from layer 5 of cortex. While this is a useful classification parameter, there are at least two major provisos.

First, often the term first or higher order is applied to histologically defined nuclei as described above. Again, as detailed in Chapter 6, it appears that first order nuclei are entirely first order, which logically follows from the observation that they receive no layer 5 input.

FIGURE 9.1 Continued

is contrasted with relay cells receiving driver input from a single source (not shown). D. Core versus matrix projections. Most of the geniculocortical projection, an example of a core system, mainly targets middle cortical layers in a highly topographical manner (*left*), whereas the projection from the central median nucleus, an exemplar of a matrix system, diffusely targets upper cortical layers, chiefly layer 1 (*right*) and often innervates multiple cortical areas (*dashed blue lines*) and/or noncortical targets such the basal ganglia and amygdala (not shown). E. Thalamocortical afferents can be either driver or modulator. This example shows the mixture of the two types (see text for details). F. Excitatory versus inhibitory patterns of thalamocortical projection. Some projections, such as the geniculocortical, chiefly innervate excitatory cells in cortex, whereas other projections, such as from the medial dorsal nucleus to the prefrontal cortex, chiefly innervate inhibitory cells (see text for details).

Abbreviations: *E or I*, excitation or inhibition; *CM*, central median nucleus; *FO or HO*, first or higher order; *GABA*, γ-aminobutyric acid; *LGN*, lateral geniculate nucleus; *MD*, medial dorsal nucleus; *PFC*, prefrontal cortex; *POm*, posterior medial nucleus; *S1 or S2*, primary or secondary somatosensory cortex; *V1 or V2*, primary or secondary visual cortex; *SpV*, spinal nucleus of the fifth nerve; *VPm*, ventral posterior medial nucleus.

Examples are the lateral geniculate nucleus, ventral division of the medial geniculate nucleus, and the ventral posterior nucleus (reviewed in Sherman and Guillery, 2013). However, nuclei defined as higher order often appear to contain both first and higher order circuitry. Thus, although the pulvinar and medial dorsal nucleus receive most driver input from layer 5 of cortex, these nuclei are also innervated by the superior colliculus with inputs that may be driver (Kelly et al., 2003; Sommer and Wurtz, 2004a,b). Second, we describe below a classification scheme that divides thalamocortical relays into core and matrix moieties, and, as noted, core and matrix projections are not closely correlated with thalamic nuclear divisions. Indeed, as noted, both first and higher order nuclei may contain both core and matrix elements.

Since the functional unit of thalamus is the relay cell, it may be useful to identify first and higher order relays at the single cell level. This would be a worthwhile classification if a thalamic relay cell received driver input only from subcortical or cortical sources and never both. More data are needed to determine if this holds true. However, even here the first and higher order classification breaks down if, say, one cortical layer 5 driver innervates a relay cell that targets middle cortical layers (a core property), while another innervates a relay cell that targets layer 1 (a matrix property). Furthermore, there is evidence that some individual relay cells receive convergent driver input from both layer 5 of cortex and a subcortical source (Bickford et al., 2015), as described below.

The functional organization of higher order relays is considered in more detail in Chapter 6.

9.1.2.2 Triadic Versus Non-Triadic Circuitry

Triadic circuitry has been described in detail in Chapter 7. We just point out here that driver inputs to thalamic relay cells can be classified as triadic or non-triadic (Figure 9.1B). This is correlated with the X/Y classification in the cat, but, given the ubiquity of triadic circuitry throughout thalamus (reviewed in Jones, 2007; Sherman and Guillery, 2013), it follows that triadic circuitry among driver inputs represents a useful classification criterion.

9.1.2.3 Mainly Single Versus Convergent Input

The classic view of thalamic function is that it operates mainly to control the flow of information to cortex without providing further processing of information (Figure 9.1C). The lateral geniculate nucleus is a good example of this. As synaptic hierarchies are ascended in retina, receptive fields become increasingly elaborated, particularly with respect to the expansion and complexity of its spatial parameters, leading, for example, to the center/surround receptive fields of many ganglion cells. The same could be said for processing in visual cortex, as receptive fields evolve from simple to complex to hypercomplex, etc. (Hubel and Wiesel, 1961, 1962, 1965, 2005). This receptive field elaboration can be seen as the integrative action of convergent driver inputs carrying different messages. The one exception to this pattern is the retinogeniculate synapse, across which virtually no spatial receptive field elaboration takes place, although there are effects of temporal structure in the relay, such as represented by the burst/tonic response modes (Sherman, 2001; Sherman and Guillery, 2013; Alitto et al., 2019). Instead, this has led to the common view that the thalamus is organized as a gated, modulatory relay rather than a site of further information processing.

If driver input from different sources were found to converge onto single relay cells, this would require an expansion of this traditional view of thalamic function and would represent a class of thalamic relay different from the conventional one exemplified by the lateral geniculate nucleus. Such evidence has recently emerged. For instance, there is evidence of convergence of direction-selective retinal inputs to cells of the lateral geniculate of the mouse to create nondirectional axis selectivity in the geniculate neurons (Marshel et al., 2012). Another study has shown that single neurons of the posterior medial nucleus of mice and rats receive convergent driver input from somatosensory cortex and the spinal nucleus of the fifth nerve (Groh et al., 2014). Another has shown in mice that, in the shell portion of the lateral geniculate nucleus, possible driver inputs from the superior colliculus and the retina converge onto single neurons (Bickford et al., 2015). In addition, evidence for layer 5 cortical input convergence onto single thalamic neurons can be inferred from functional studies involving prefrontal inputs onto the medial dorsal thalamus, where thalamic neurons represent conjunctions of cortical signals (Rikhye et al., 2018), and a recent electron microscope study of mice has demonstrated convergence of layer 5 inputs from primary somatosensory and motor cortices onto single thalamic neurons in the posterior medial nucleus (Sampathkumar et al., 2021). These are important results, but we require more examples and larger samples to determine the extent to which such a new category of thalamocortical circuitry exists.

9.1.3 Classification Based on Thalamocortical Projections

Both morphological and functional criteria have been used to classify thalamocortical projection patterns, but, as we demonstrate here, these are neither complete nor mutually exclusive.

9.1.3.1 Core and Matrix

The concept of a "reticular activating system," meaning a rather diffuse, nonspecific brainstem to thalamus to cortex circuit involved in arousal and sleep-wakefulness, has been with us since at least the 1950s (Moruzzi and Magoun, 1949; Magoun, 1952; Garcia-Rill, 2009). This has been contrasted with a different thalamocortical system seen as specific and topographic that, for example, underlies sensory processing at the cortical level (Figure 9.1D).

Jones (2001) has more recently revisited this issue with a proposal that thalamocortical projections can be classified into two components that he called "core" and "matrix." According to the core/matrix classification, the matrix moiety involves far-reaching and diffuse thalamocortical projections that target upper layers, particularly layer 1, and these relay cells can also target noncortical areas such as the basal ganglia and amygdala. Whether these cortical and noncortical projections emanate from different thalamic cell populations or branching axons of single neurons remains to be determined. In this regard, the matrix would appear simply to be a new designation for the reticular activating system. Core projections, in contrast, are highly topographic and target mainly middle layers, particularly layer 4, and these would appear to represent what had been thought of as specific thalamocortical systems.

Jones (2001) based the core/matrix classification on histological material from monkeys, in which the relevant thalamic relay cells express reactivity for different calcium binding proteins: the smaller calbindin-positive cells represent the matrix class, and the larger parvalbumin-positive cells the core class. While this distinction based on calcium binding proteins may work for the monkey, its use as a universal identification tool across mammalian species for this classification seems doubtful (Ichida et al., 2000; Amadeo et al., 2001; Nakamura et al., 2015).

The basic idea behind this classification is that the core projections bring specific and topographic sensory information to cortex, and the matrix projections serve a diffusely organized role in recruitment of more widespread thalamocortical ensembles to reflect changes in behavioral state, such as arousal or attentional shifts, and possibly to play a role in "binding" discharge patterns of cortical cells that represent different parts of a perceptual whole (Singer and Gray, 1995; Fries, 2009). This is not fundamentally different from the 1950s idea that the reticular activating system (read: matrix) is involved in overall behavioral effects linked to levels of arousal and that the specific, topographic thalamocortical systems (read: core) are used for specific processing of the environment.

As noted, this core/matrix classification does not map well onto thalamic nuclear divisions. Although some nuclei, such as the midline nuclei, seem entirely, or nearly so, comprised of matrix projections, matrix circuits also seem to exist in most other nuclei, even those considered to be mostly core. The lateral geniculate nucleus of the monkey serves as an example of a problem with this classification and its presumed functional interpretation as applied to entire nuclei. By all criteria, the parvalbumin positive parvocellular and magnocellular cells seem to represent core projections, but the koniocellular cells seem more ambiguous. These koniocellular cells are calbindin-positive and thus should be part of the thalamocortical matrix, but they are not noticeably more diffusely organized than are parvocellular and magnocellular cells. Koniocellular receptive fields, while somewhat larger, are nonetheless well-focused and constrained to a small part of visual field, and their terminations in cortex are generally well-localized in upper layers, including layer 1 (Irvin et al., 1986, 1993; Norton et al., 1988; Ding and Casagrande, 1998; Casagrande and Xu, 2004; Roy et al., 2009; Cheong et al., 2013). This projection to layer 1 is often used to place at least some koniocellular geniculate cells on the matrix team, but the lack of clear diffuseness in their terminations in layer 1 raises concerns with this classification.

Perhaps the biggest problem with the core/matrix scheme is evidence that branching patterns of many thalamocortical neurons fail to fit into either classification (Clasca et al., 2012; Kuramoto et al., 2017). It seems clear that a proper classification scheme based just on branching patterns of thalamocortical arbors would result in many more than two categories.

9.1.3.2 Different Laminar and Areal Targets of Thalamocortical Projections

The core/matrix division of thalamocortical projections can be viewed as a special case of differences based on laminar terminations or combinations of cortical areas targeted by individual axons (Figure 9.1D). Indeed, many variations exist regarding laminar terminations among these arbors, and, whereas Figure 9.1D suggests that only matrix arbors branch to innervate multiple cortical areas, examples of core projections also do so. Geniculate X and Y cells in the cat exemplify both features (see below discussion and Figure 9.1B). X axons target the lower half of layer 4 of V1 alone, whereas Y arbors target the upper half of layer 4 in V1[1] and also branch to innervate V2.

9.1.3.3 Drivers and Modulators

It has been clear for some time that glutamatergic inputs in both thalamus and cortex, including thalamocortical input, can be classified as driver or modulator (Figure 9.1E). This has been described in some detail in Chapter 5. There is evidence in mice that thalamocortical projections contain both types. Examples of such thalamocortical driver inputs so far include the following:

- Regarding inputs from the ventral posterior nucleus to the primary somatosensory cortex, all thalamocortical inputs to cells in layers 4–6 and about 25% to cells in layers 2/3 are driver (Viaene et al., 2011b,c). The same pattern exists for inputs from the ventral division of the medial geniculate nucleus to primary auditory cortex (Viaene et al., 2011b, 2011c). The pattern for geniculocortical projections has yet to be investigated in full, but cross-correlation data from the cat strongly suggests that geniculocortical input to layer 4 cells is driver (Reid and Alonso, 1995; Usrey et al., 2000; Alonso et al., 2001).
- The projection from the posterior medial nucleus to cells in layer 4 of the second somatosensory area is driver (Viaene et al., 2011a), and the equivalent projection from the dorsal division of the medial geniculate nucleus to the secondary auditory area is also driver (Lee and Sherman, 2008).
- The projection of the posterior medial nucleus to cells in layers 2–6 of the primary motor cortex is all driver (Mo and Sherman, 2019).

Examples of such thalamocortical modulator inputs so far include the following:

- About 75% of the inputs from the ventral division of the medial geniculate nucleus and ventral posterior nucleus to cells in layers 2/3 of the primary auditory and somatosensory cortices, respectively, are modulator (Viaene et al., 2011b).
- The projection from the posterior medial nucleus to cells in layers 2–6 of the primary somatosensory nucleus is all modulator (Viaene et al., 2011a).

The significant representation of modulator input among thalamocortical projections is rather surprising and suggests that these thalamocortical inputs cannot all be lumped together as information-bearing. A particularly interesting example is the prominent modulation suggested for the projection from the higher order posterior medial nucleus to the primary somatosensory cortex. An outstanding question is what drives these relay cells. For two examples: If they are driven by layer 5 cells of the secondary somatosensory cortex, this would represent a transthalamic feedback modulatory pathway from the secondary to the primary somatosensory cortex; if instead these cells are driven by layer 5 cells of the primary

---

1. It should be noted that, in the cat, V1 is usually referred to as "area 17" and V2 as "area 18." We shall continue to use V1 and V2 nomenclature here to be more consistent among the different species we describe.

somatosensory cortex, this would represent a transthalamic feedback modulatory pathway from the primary somatosensory cortex back to itself. Evidence for both possibilities exist (Miller and Sherman, 2019). Many more potentially important but unresolved thalamocortical patterns of driver versus modulator likely exist, and so clearly more examples are needed to determine any overall pattern of drivers and modulators among thalamocortical inputs.

9.1.3.4 Excitatory Versus Inhibitory Cortical Targets

Given that cortex has a large variety of different cell types, it follows that thalamocortical motifs may be classified based on differences in the mix of such cortical cells targeted by any thalamocortical input (Figure 9.1F). While the possibility of such heterogeneity seems quite rich, so far the main attempts to distinguish postsynaptic partners of thalamocortical inputs relates to patterns of excitatory or inhibitory neuronal targets. For example, thalamocortical inputs from first order sensory thalamic nuclei contact both excitatory and inhibitory cortical neurons, and the inhibitory targets are predominantly parvalbumin (PV)-positive interneurons (Tuncdemir et al., 2016). However, projections from higher order nuclei, such as the posterior medial nucleus, also target vasoactive intestinal peptide (VIP)–positive interneurons (Williams and Holtmaat, 2019), and these appear to be involved in disinhibitory control (Pi et al., 2013). Finally, functional evidence directed at projections from the medial dorsal nucleus to prefrontal cortex suggests that these projections are far more efficient at driving cortical inhibition compared to the inhibitory impact of a first order sensory thalamic input on its cortical targets (Figure 9.1F) (Schmitt et al., 2017).

A recent finding in the mouse indicates that projections to visual cortex from the lateral geniculate nucleus (a first order nucleus) chiefly target excitatory neurons, whereas those from the medial dorsal nucleus (a higher order nucleus) to prefrontal cortex chiefly target inhibitory neurons (Li et al., 2020). This might suggest a different motif for first versus higher order thalamocortical circuitry. However, another recent study showed that the projection from the posterior medial nucleus to the second somatosensory cortex in the mouse (a higher order projection) predominantly targets excitatory neurons (Sampathkumar et al., 2021), and this argues against such an interpretation.

9.1.4 Classification Based on Parallel Thalamocortical Streams

Parallel processing streams to classify thalamocortical pathways may be distinguished in terms of differences in both driver inputs and thalamocortical outputs. This may accordingly be seen as a particularly useful means of identifying thalamocortical types because most or all such identified classes likely function independently and in parallel with respect to one another. Unfortunately, this classification approach has been used successfully only with respect to the lateral geniculate nucleus and mainly in carnivores and primates. We describe the organization of these streams below, and then revisit them in Chapter 11 for a more thorough examination of their physiological properties.

9.1.4.1 W, X, and Y Pathways in the Cat

The best known of these parallel neuronal streams are the W, X, and Y pathways in the cat (for details, see Sherman and Spear, 1982; Stone, 1983; Sherman, 1985; see also Figure 9.1B;

W pathway not shown in figure).[2] These each start with their own class of retinal ganglion cell that provides the driver input to the corresponding class of geniculate relay cell, and each of these classes has separate projections to cortex. This separation of inputs and outputs is especially clear for the X and Y streams (Figure 9.1B), as already noted (Wilson et al., 1984; Humphrey et al., 1985a,b; Hamos et al., 1987). On the input side, X circuitry is dominated by driver inputs operating through triadic circuitry, which is generally missing for the Y pathway; on the output side, the X stream terminates solely in V1 in single patches in lower layer 4, whereas the Y pathway terminates in multiple patches in upper layer 4 in V1 and also terminates in V2 via branching axons.

Quantitative differences between the X and Y streams regarding their relative distribution at different levels of the visual neuraxis bear mentioning (reviewed in Sherman, 1985). In the cat's retina, the X to Y ratio among ganglion cells is roughly 10:1 (Wässle, 1982), but the ratio is much closer to unity in the lateral geniculate nucleus (Friedlander et al., 1981). The difference in ratios between retina and thalamus is likely related to the observation that retinogeniculate Y arbors, compared to X arbors, are much larger with more synaptic terminals and thus appear to innervate more geniculate relay cells (Sur et al., 1987). A similar difference exists among geniculocortical arbors (Humphrey et al., 1985a,b): Y arbors are larger, innervate more cortical territory, and support more synaptic terminals. Thus, whereas X cells dominate Y cells numerically in the retina, the opposite seems true for the relative influence of the X and Y streams in cortex.

An interpretation of these observations can be found elsewhere (Sherman, 1985). Briefly, the suggestion is that the Y system is more important than is the X system for cat vision. The Y system provides basic luminous contrast sensitivity for the animal, whereas the X system complements this by providing such features as better acuity. In the cat, surgical removal of V1, which removes all X but not all Y representation in cortex, lowers the animal's acuity somewhat but otherwise leaves it with excellent vision, suggesting that even this limited Y representation at the cortical level is adequate for subserving much basic vision (Lehmkuhle et al., 1984). Because Y cells code lower spatial frequencies than do X cells, complete retinal tiling of each type leads to a low Y to X cell ratio, but the greater importance to the cat of Y cells and their responsiveness to lower spatial frequencies leads to a high Y to X cell ratio in terms of cortical representation.

The W pathway is less well studied and the phrase "W cells" seems to represent a catchall of several separate populations best defined as neither X nor Y. The relationship of this input to triadic circuitry in the lateral geniculate nucleus is unknown, but, on the output side, it terminates chiefly in layers 2/3 within V1 and also innervates multiple extrastriate cortical areas (Sherman and Spear, 1982; Stone, 1983; Sherman, 1985). Whether this latter pattern is due to branching axons, separate neuronal populations, or some combination thereof remains unknown.

---

2. There has been some terminological variability for these classes. For instance, cat W, X, and Y have been called "sluggish," "sustained," and "transient," respectively (Cleland et al., 1971b, 1976), and the "sluggish" category has been further subdivided with exotic names (Cleland and Levick, 1974). Old and curiously emotional arguments about such terminological debates exist (Rowe and Stone, 1977; Hughes, 1979). A morphological approach to retinal ganglion cells has termed X cells "beta," Y cells, "alpha," and W cells, "gamma" (Boycott and Wässle, 1974; Wässle, 1982), with other types given names representing further progression up the Greek alphabet (e.g., (Berson et al., 1998, 1999). In any case, the "W/X/Y" terminology has emerged as the current terminological choice.

9.1.4.2 K, P, and M Pathways in the Monkey

Available data from the monkey indicate an arrangement of parallel retino-geniculo-cortical streams that is quite similar, even homologous, to those in the cat, although monkey pathways have been much less studied. As noted in Chapter 11, the monkey pathways are called the koniocellular, magnocellular, and parvocellular pathways, which appear homologous to the W, Y, and X pathways, respectively (Casagrande and Xu, 2004). Similar to the differences in cortical projection patterns in the cat, magnocellular and parvocellular geniculate neurons chiefly target layer 4, with magnocellular axons terminating dorsal to the parvocellular axons and producing larger terminal arbors, and the koniocellular pathway seems to involve multiple classes that mainly innervate layers 2/3 (Figure 9.2; Ding and Casagrande, 1997; Martin et al., 1997; White et al., 2001; Casagrande and Xu, 2004).

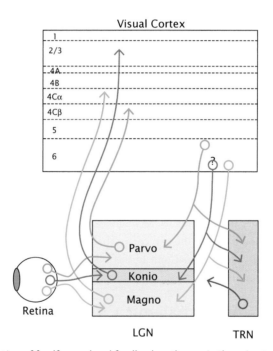

FIGURE 9.2 Organization of feedforward and feedback pathways in the primate. Three major parallel processing streams are established in the retina: the parvocellular, magnocellular, and koniocellular streams, indicated schematically with green, blue, and red cell bodies and axons. Retinal ganglion cells belonging to these streams send axons to distinct layers of the lateral geniculate nucleus to synapse with relay neurons that selectively innervate cortical layers 4Cα, 4Cβ, and layers 2 and 3. Although not indicated, relay neurons also often provide sparse input to layers 1 and 6. Layer 6 corticogeniculate neurons are also organized into three major streams. Neurons in upper layer 6 selectively target the parvocellular layers of the lateral geniculate nucleus, while neurons in lower layer 6 target the magnocellular and koniocellular layers (note: although it is unclear in primates whether separate neurons in lower layer 6 target the magnocellular and koniocellular layers, physiological results from monkeys and anatomical evidence from tree shrews suggests separate populations (Briggs and Usrey, 2009; Usrey and Fitzpatrick, 1996). In addition to making synapses with neurons in the lateral geniculate nucleus, corticogeniculate neurons also send axon collaterals into the thalamic reticular nucleus to synapse with GABAergic neurons that, in turn, project to the lateral geniculate nucleus.

9.1.4.3 Other Evidence for Parallel Streams Through Thalamus

In vision, parallel streams like those just described have been documented in all other mammalian species so far studied for this property, including tree shrews, ferrets, minks, and bushbabies (Stone, 1983; Casagrande and Xu, 2004). Unfortunately, a clear demonstration of such parallel processing in mice and rats remains to be done (but see Krahe et al., 2011; Denman and Contreras, 2016).

The generality of parallel processing through thalamus, especially outside the lateral geniculate nucleus, remains mainly a hypothesis to be tested. Limited anatomical data suggest parallel processing in the somatosensory thalamus of the cat (Yen and Jones, 1983). Otherwise, given that triadic circuitry is a distinguishing feature for the separate X and Y parallel streams in the cat (see above discussion), and given the ubiquity of triads throughout the mammalian thalamus, it seems likely that similar parallel streams identified in terms of triadic or non-triadic circuitry exist widely. As is so often the case, more data are needed to establish the extent to which parallel processing is a general feature in the classification of thalamocortical motifs (Halassa and Sherman, 2019).

9.2 Corticothalamic Motifs

A subset of pyramidal cells in cortical layers 5 and 6 have axons that leave the cortex to innervate the thalamus. Because cells in layers 5 and 6 are the only cells in the cortex with axons that target subcortical structures, they play a critical role in allowing the cortex to interact with other parts of the neuraxis and influence behavior. Layer 6 cells provide modulatory input to first order and higher order thalamic nuclei. In contrast, layer 5 cells provide driving input only to higher order thalamic nuclei. As might be expected from their different synaptic properties and projection patterns, the two corticothalamic pathways operate independently of each other and serve different functions. In the following sections, we examine these differences and speculate on their functional contributions to neural processing and behavior.

9.2.1 Layer 6 Corticothalamic Circuits

9.2.1.1 General Features of Layer 6 Corticothalamic Circuits

The majority of layer 6 neurons with axons that leave the cortex and project to the thalamus target thalamic nuclei that provide the cortical area with feedforward input.[3] These layer 6 neurons appear to provide feedback modulatory influence over thalamus. For example, layer 6 corticothalamic cells in visual cortex have axons that target the lateral geniculate nucleus which contains relay cells that receive feedforward signals from the retina and project to

---

3. A subset of layer 6 neurons with axons that leave the cortex project to the claustrum and not to the thalamus. Interestingly, the regions of the claustrum innervated by these layer 6 cells contain neurons that project back to the same cortical area, thereby forming a corticoclaustral loop for information exchange (LeVay and Sherk, 1981; LeVay, 1986; Katz, 1987).

the visual cortex (reviewed in Briggs and Usrey, 2008; Sherman and Guillery, 2013). Layer 6 corticothalamic neurons also give rise to local axon collaterals that terminate in the layers of cortex, especially layer 4, that receive thalamic input (Lund and Boothe, 1975; Gilbert and Wiesel, 1979; Usrey and Fitzpatrick, 1996; Briggs et al., 2016). Layer 6 neurons are therefore in a strategic position to influence thalamocortical communication since they target both the origin of the thalamic input as well as its cortical targets.

Importantly, layer 6 corticothalamic neurons in different cortical areas align with different thalamic nuclei. For instance, whereas layer 6 neurons in visual cortex send feedback axons to the lateral geniculate nucleus, layer 6 neurons in somatosensory cortex send axons to the ventral posterior nucleus, which receives feedforward input from the head and body via the medial lemniscus, and layer 6 neurons in auditory cortex send feedback axons to the ventral division of the medial geniculate nucleus, which receives feedforward auditory signals from the inferior colliculus. Because the synaptic properties of layer 6 input to thalamus are largely similar across thalamic nuclei, in the following sections we will use the feedback pathway from visual cortex to the lateral geniculate nucleus as a model system for understanding corticothalamic feedback.

Numerically, the corticothalamic feedback pathway is robust, with layer 6 axons providing 30–50% of all synapses onto relay cells (Van Horn et al., 2000). However, compared to feedforward inputs that provide large synapses onto proximal regions of relay cell dendrites, the synapses from corticothalamic axons are relatively small and are located more distally from the cell body (Wilson et al., 1984; Erisir et al., 1997a). These features, along with the relatively small size of corticothalamic EPSPs compared to feedforward EPSPs and postsynaptic activation of metabotropic glutamate receptors not seen with the feedforward input, establish the modulatory, rather than driving, influence that layer 6 projections have on thalamic cells (see Chapter 5; reviewed in Sherman and Guillery, 1998, 2013).

9.2.1.2 Parallel Streams in the Corticogeniculate Feedback Pathway

There is a striking relationship between the response properties and projections of corticogeniculate neurons and the parallel processing streams that organize the geniculate relay cells they innervate. For instance, and as described in greater detail in Chapter 11, the geniculocortical pathway in primates is comprised of the magnocellular, parvocellular, and koniocellular streams (Figure 9.2; reviewed in Usrey and Alitto, 2015). These streams are established by separate populations of retinal ganglion cells that innervate specific cell types in different laminae of the lateral geniculate nucleus that, in turn, project to distinct layers of primary visual cortex. The axons of the magnocellular neurons conduct action potentials more rapidly than those of the parvocellular neurons, which, in turn, are faster conducting than those of the koniocellular neurons. With respect to their visual physiology, magnocellular neurons are more sensitive to luminance contrast, whereas parvocellular cells show wavelength discrimination[4] but reduced luminance contrast sensitivity. Compared to parvocellular

---

4. Wavelength discrimination for parvocellular cells is found in monkeys active during the day, like rhesus monkeys, but not in monkeys active at night, like owl monkeys (Jacobs et al., 1993, 1996; Levenson et al., 2007).

neurons, magnocellular neurons have larger receptive fields and greater extraclassical surround suppression, a phenomenon whereby stimuli outside a cell's classical receptive field suppress responses to stimuli within the classical receptive field (Alitto and Usrey, 2008; Archer et al., 2021). Magnocellular neurons also produce transient responses to visual stimuli, whereas the responses of parvocellular neurons are more sustained. The axons of magnocellular and parvocellular neurons in the lateral geniculate nucleus terminate in layer 4 of visual cortex, layers 4Cα and 4Cβ, respectively. In contrast, koniocellular neurons have axons that bypass layer 4C and terminate in layers 1–3. Compared to magnocellular and parvocellular neurons, much less is known about the physiological properties of koniocellular neurons.

Corticogeniclate feedback in the monkey is comprised of stream-specific projections, and these projections show a correspondence with the parallel feedforward projections (Figure 9.2). Namely, there are distinct classes of corticogeniculate neurons and their axons selectively target the magnocellular, parvocellular, and perhaps even the koniocellular layers of the lateral geniculate nucleus, with parvocellular projecting neurons occupying the upper third of layer 6 and magnocellular and koniocellular projecting neurons occupying the lower third (Fitzpatrick et al., 1994; Ichida and Casagrande, 2002; Ichida et al., 2014). It is worth noting that neurons projecting to the koniocellular-like layers in other species also have cell bodies located in the very bottom of layer 6, and cells in this region of layer 6 also project to the pulvinar (Usrey and Fitzpatrick, 1996; Hoerder-Suabedissen et al., 2018).

While the idea of parallel streams of corticogeniculate feedback is supported by evidence from other species, including the rat, cat, ferret, and tree shrew (Tsumoto and Suda, 1980; Bourassa and Deschênes, 1995; Briggs and Usrey, 2005; Hasse et al., 2019), our understanding of the physiological properties of corticogeniculate neurons and the relationship between anatomy and physiology is perhaps greatest in the monkey. Specifically, corticogeniculate neurons in the lower region of layer 6, where cells with projections to the magnocellular layers of the lateral geniculate nucleus are located (Fitzpatrick et al., 1994), have fast conducting axons, show robust extraclassical surround suppression, and are more responsive to luminance contrast and fast-moving stimuli, which are properties of the feedforward magnocellular stream (Briggs and Usrey, 2009). In contrast, corticogeniculate neurons in the upper region of layer 6, where cells targeting the parvocellular layers are located (Fitzpatrick et al., 1994), have slower conducting axons, have modest extraclassical suppression, and are less responsive to luminance contrast and fast-moving stimuli, which are properties of the feedforward parvocellular stream (Briggs and Usrey, 2009). As a result of this organization, visual stimuli well-suited for driving the magnocellular pathway should preferentially excite layer 6 feedback cells that innervate the magnocellular layers of the lateral geniculate nucleus, while visual stimuli better suited for exciting the parvocellular pathway should differentially excite layer 6 neurons that innervate the parvocellular layers. In other words, corticogeniculate neurons are positioned to modulate visual signals traveling from the thalamus to the cortex in a stream-specific fashion.

A third population of corticogeniculate neurons in the monkey also located in the bottom of layer 6 where magnocellular projecting neurons are found (Fitzpatrick et al., 1994; see also Usrey and Fitzpatrick, 1996) share many features in common with the feedforward koniocellular stream (Briggs and Usrey, 2009). Although less is known about koniocellular geniculate neurons compared to their magnocellular and parvocellular counterparts, existing evidence indicates that koniocellular geniculate neurons have slower conducting axons and slower visual responses compared to magnocellular and parvocellular neurons and contrast sensitivity and temporal frequency tuning that is intermediate to magnocellular and parvocellular neurons (Irvin et al., 1986; Norton et al., 1988; White et al., 1998; Solomon et al., 1999; Hendry and Reid, 2000). A particularly noteworthy trait of koniocellular geniculate neurons is the prominence of input from retinal short-wavelength cones (reviewed in Hendry and Reid, 2000; Solomon and Lennie, 2007). All of these traits are shared with corticogeniculate neurons that have axons presumed to innervate the koniocellular geniculate layers. Taken together, it is rather remarkable how closely the physiology of the three classes of corticogeniculate neurons aligns with the feedforward magnocellular, parvocellular, and koniocellular streams.

9.2.1.3 Excitatory and Inhibitory Influence of Corticothalamic Feedback: Cellular Targets

Across species, corticogeniculate axons target excitatory relay cells and GABAergic interneurons in the lateral geniculate nucleus as well as GABAergic neurons in the neighboring thalamic reticular nucleus that, in turn, innervate geniculate relay cells (reviewed in Sherman and Guillery, 2013; Usrey and Alitto, 2015). Corticogeniculate neurons therefore have the opportunity to modulate relay cell activity via direct excitation and disynaptic inhibition. Various patterns of this feedback are discussed in Chapter 11 (see Figure 7.1B). Whether or not all corticogeniculate neurons innervate both geniculate relay cells plus reticular cells and/or interneurons is an open and unresolved question. These are important questions to answer since synapses between corticogeniculate neurons and their target neurons in the lateral geniculate nucleus and the reticular nucleus show varying degrees of rate-dependent facilitation and depression (Crandall et al., 2015).

As will be discussed in greater detail in Chapter 11, a topographically ordered center/surround organization of corticogeniculate projections has implications for effects involving cognitive activities. For instance, a center/surround organization of net excitation and suppression from feedback as shown in Figure 7.1C could provide a substrate for a "spotlight" of attentional modulation of thalamic activity that could serve to shift geniculate responses between burst and tonic response modes as well as modulate overall activity levels. Along these lines, spatial attention has been shown to: (1) decrease visual responses in the thalamic reticular nucleus (McAlonan et al., 2006); (2) increase activity in the lateral geniculate nucleus (McAlonan et al., 2006); and (3) increase the strength of synaptic transmission between neurons in the lateral geniculate nucleus and visual cortex (Briggs et al., 2013). Evidence also indicates there is a relationship between cortical state and visual activity in the lateral geniculate nucleus (McGinley et al., 2015). For instance, studies examining network interactions reveal oscillatory feedback interactions between cortex and the lateral geniculate nucleus in the α frequency band (Bastos et al., 2014). These interactions could serve to synchronize activity between geniculate cells, which is predicted to augment communication between thalamus and cortex. These topics will be discussed more thoroughly in Chapter 10.

9.2.2 Layer 5 Corticothalamic Circuits

9.2.2.1 Diversity of Layer 5 Corticofugal Projections

Pyramidal neurons in cortical layer 5 innervate numerous subcortical targets, including thalamus, basal ganglia, various brainstem sites, and, in some cases, spinal cord. Individual axons of these cells branch repeatedly to innervate many or all of these targets (see below). These projections are not unique to a subset of cortical areas; indeed, every cortical area for which appropriate information is available produces such a corticofugal output. Importantly, it is via these projections that the cortex can communicate with other parts of the neuraxis and thus affect behavior.

There appear to be two main groups of layer 5 pyramidal cells: one projects subcortically and the other projects to other cortical areas (Petrof et al., 2012). For both groups, projecting axons branch to innervate their local cortical regions in addition to their distant targets. Features that distinguish the two groups include (reviewed in Sherman and Guillery, 2013; Sherman, 2014, 2016) (1) the subcortically projecting cells are located more ventrally in layer 5 (often referred to as layer 5b) than are the corticocortically projecting cells (said to be in layer 5a); (2) the subcortically projecting cells are the largest pyramidal cells in cortex and typically have apical dendrites reaching all the way to layer 1, whereas the other pyramidal cells in layer 5 are smaller and typically have shorter apical dendrites that fail to reach layer 1; and (3) the subcortically projecting layer 5 cells often fire in burst mode based on the activation of a voltage- and time-dependent Ca^{2+} conductance, a firing mode not seen in the other layer 5 cells (Larkum et al., 1999; Llano and Sherman, 2009). Having distinguished these two groups of layer 5 cells, the remainder of this chapter is limited to those that project subcortically.

9.2.2.2 Thalamic Innervation

Layer 5 axons targeting higher order thalamic nuclei innervate relay cells on their proximal dendrites, often in glomeruli (see Chapter 7 for description of glomeruli) in triadic arrangements, whereas other inputs are typically found outside glomeruli in simple contacts onto dendrites (reviewed in Jones, 2007; Sherman and Guillery, 2013). Overall, only about 2% of all synapses made with higher order relay cells come from layer 5 input (Wang et al., 2002; Van Horn and Sherman, 2007).

As noted above, all thalamic relay cells appear to receive input from layer 6 cells, an input organized mainly but not exclusively in a feedback manner. In addition, some relay cells receive an input from layer 5 of cortex that is rarely organized in a feedback manner, meaning that the cortical area of origin is usually not a target of the postsynaptic relay cell. An example of such circuitry is shown in Figure 6.1. In this example, the layer 5 projection is feedforward, which is to say that it is organized to ascend a hierarchical pattern of thalamo-cortical processing. Also, in contrast to the modulatory action of the layer 6 corticothalamic projection, the layer 5 projection is driver. Thus, higher order thalamic nuclei that receive driving signals from one cortical area and relay those signals to another provide a means for cortico-cortical communication that is distinct from any direct connections between cortical areas. This role for higher order nuclei is discussed in greater detail in Chapter 6. Last, in addition to providing driving input to cortex, higher order thalamic nuclei also provide modulatory input to cortical areas that appears to align with the criteria that distinguish feedback projections, as discussed in Chapter 5.

9.2.2.3 Extrathalamic Innervation

There are additional noteworthy differences between layer 5 and layer 6 corticothalamic neurons. Layer 6 corticothalamic axons effectively innervate only thalamus,[5] including the thalamic reticular nucleus with branching arbors there (Sherman and Guillery, 2013). Layer 5 axons also innervate thalamus and pass through the thalamic reticular nucleus, providing what appear to be terminals *en passant* there (Prasad et al., 2020). Layer 5 axons branch further to innervate a number of other subcortical sites (reviewed in Sherman and Guillery, 2013; Sherman, 2016). Chapter 13 further elaborates branching of layer 5 corticofugal axons.

9.3 Summary and Concluding Remarks

We have identified numerous criteria that can be used to classify thalamocortical and corticothalamic motifs, and we make no claim that the list is exhaustive. This promotes a significant challenge because the criteria listed are pretty much independent of one another, which means they could be combined in multiple ways to produce a rather large number of unique classes. For instance, core thalamocortical projections can be further classified based on the pattern of cortical layers innervated, whether their driver inputs are triadic, whether their driver inputs are significantly convergent, whether their driver inputs are subcortical or from layer 5 of cortex, whether their outputs are driver or modulator, whether multiple cortical areas are targeted via branching axons, etc.

This challenge, large as it is, should not deflect from the need for such a classification. Until we have a complete classification of thalamocortical and corticothalamic motifs, we cannot fully grasp the heterogeneity in the functional connections between thalamus and cortex, and this remains a limitation to our understanding of overall cortical functioning. Furthermore, there are numerous neurological and psychiatric disorders associated with thalamic pathology (e.g., Means et al., 1974; Rafal and Posner, 1987; Danos et al., 2003; Brickman et al., 2004; Chauveau et al., 2005; Byne et al., 2009; Cronenwett and Csernansky, 2010; Zwergal et al., 2011; Janssen et al., 2012; Parnaudeau et al., 2013; Tu et al., 2018). Knowledge of which motifs are involved in each case could plausibly assist in improvements in treatment and/or prevention.

We have emphasized the many attempts to classify thalamocortical and corticothalamic projection types. This implicitly underscores the need for such a classification, an implicit need that we wish to make clearly and explicitly (Halassa and Sherman, 2019). Yet each attempt currently to do so is seriously flawed or limited. The methodological tools to develop such a complete classification that we suggest is necessary are probably not yet available. However, the speed of technological advance in neuroscience is so impressive that such

---

5. An exception to this is evidence that some layer 6 cells from auditory cortex project to innervate the inferior colliculus (Kunzle, 1995; Coomes et al., 2005; Bajo et al., 2007; Slater et al., 2013).

tools may soon be accessible. It is in this context that we argue that identifying the importance of a thalamocortical classification, as we have attempted to do here, is a useful step in the process.

9.4 Some Outstanding Questions

1. Do individual neurons in the thalamic reticular nucleus receive stream-specific or mixed-stream input from layer 6 feedback neurons? Likewise, are the projections of thalamic reticular neurons specific for cell classes in the lateral geniculate nucleus?
2. How does layer 6 feedback influence thalamic activity at the population level? For example, does layer 6 feedback influence rhythmic/oscillatory activity between neurons?
3. Given that individual layer 6 axons branch to target the thalamus and the cortical layers that receive thalamic input, what are the effects of the input to layer 4 on thalamocortical communication?
4. Is layer 6 feedback a route for cognitive processes to influence the thalamus?
5. Are layer 6 feedback projections to higher order thalamic nuclei comprised of parallel streams similar to those found in the projections to first order thalamic nuclei?

10

Spike Timing and Thalamocortical Interactions

The receptive field properties of sensory neurons in the thalamus and cortex have been studied for decades; however, relatively few studies have examined how individual neurons and small networks of neurons communicate with each other. This chapter explores the role of spike timing of afferent inputs to thalamus in establishing different patterns of thalamic activity as well as how spike timing affects thalamocortical communication. Because most experiments have been conducted using the visual pathway of the cat, the cat retino-geniculo-cortical pathway will be emphasized; however, details from other species and systems are included to augment the discussion.

As described in previous chapters, neurons in the lateral geniculate nucleus receive driving input from retinal ganglion cells—the output cells of the retina—and, in turn, project to neurons in primary visual cortex. Although the size and shape of the lateral geniculate nucleus vary tremendously across species (see Figure 11.2), one property appears common to all nuclei: lateral geniculate neurons have receptive fields that are extremely similar to those of their retinal afferents (Hubel and Wiesel, 1961; Levick et al., 1972; Usrey et al., 1999). Even the name given to lateral geniculate projection neurons—relay cells—implies that they do little more than pass the baton of visual activity from retina to visual cortex. Given the similarity of receptive field structure between the retina and lateral geniculate nucleus, the question often arises: What purpose does the lateral geniculate nucleus serve? Or, why not simply have a direct pathway from retina to visual cortex? A partial answer to this question can be found by examining the activity patterns of monosynaptically connected retinal ganglion cells and lateral geniculate neurons. While receptive fields do not change dramatically between the retina and the lateral geniculate nucleus, neurons in the lateral geniculate nucleus can and do transform the temporal structure of the spike trains they receive from the retina. This chapter includes and builds on previous discussions on the range of lateral geniculate responses to retinal input and the consequences of these activity patterns for driving cortical responses (Sherman, 1996, 2001; Usrey and Reid, 1999; Usrey, 2002a,b; Alitto et al., 2019; Halassa and Sherman, 2019; Usrey and Sherman, 2019).

10.1 Dynamics of Retinogeniculate Communication

Before discussing the role of spike timing for visual processing in the lateral geniculate nucleus, it is worth noting three properties of retinal and lateral geniculate activity patterns. First, with the exceptions of sleep, epilepsy, and some features of bursts (more details on bursts are described below in sections 10.3 and 10.4) essentially every spike produced by a geniculate relay cell is evoked by a retinal spike (Kaplan and Shapley, 1984; Sincich et al., 2007; Alitto et al., 2019). Second, retinal ganglion cells typically fire action potentials at higher rates than their lateral geniculate target cells (Hubel and Wiesel, 1961; Levick et al., 1972; Kaplan et al., 1987; Usrey et al., 1999; Sincich et al., 2007; Rathbun et al., 2010; Alitto et al., 2019). Thus, not all retinal spikes elicit lateral geniculate action potentials. Third, the latency of visual responses in the lateral geniculate is not simply the latency of retinal ganglion cell responses plus the delay imposed by the axon conduction velocity and synaptic transmission. Instead, it is a value greater than the sum of these events and includes the additional time required for temporal summation of retinal inputs. This third point is evident from simultaneous recordings of synaptically connected retinal ganglion cells and lateral geniculate neurons (Mastronarde, 1987; Usrey et al., 1999). As shown with a pair of on-center Y cells in Figure 10.1, even though it takes about 2.5 msec for a retinal spike to travel to the lateral geniculate nucleus and possibly trigger a postsynaptic spike, the difference in visual response latency between the retinal ganglion cell and lateral geniculate neuron is approximately 10 msec. This timing difference suggests that the initial response of a retinal ganglion cell to a visual stimulus is not sufficient to evoke a postsynaptic action potential. Rather, the spiking activity of neurons in the lateral geniculate nucleus depends on temporal summation and the history of afferent activity from the retina. Last, delayed inhibition following retinal EPSPs to lateral geniculate neurons serves many roles, including improving the temporal precision of responses in the lateral geniculate nucleus (Butts et al., 2011, 2016).

Several studies in the cat and monkey have examined the relationship between retinal spike history and lateral geniculate spike production by looking at the effect of retinal interspike interval on the probability that a retinal spike will elicit a lateral geniculate spike (Mastronarde, 1987; Usrey et al., 1998; Levine and Cleland, 2001; Rowe and Fischer, 2001; Carandini et al., 2007). These studies all conclude that retinal spikes are significantly more likely (~4–6 times more likely; Figure 10.2A) to evoke lateral geniculate spikes when they occur following the shortest interspike intervals (limited by the retinal cell's refractory period). The increased efficacy of retinal spikes then declines as interspike intervals increase to about 30 msec (Rowe and Fischer, 2001; Carandini et al., 2007). The efficacy of retinal spikes remains low for longer interspike intervals until a second phase of enhanced efficacy is seen for spikes following interspike intervals longer than 80 msec (Figure 10.2A; Alitto et al., 2019). The mechanisms that underlie this second phase of enhanced efficacy likely include a release of synaptic depression (Swadlow and Gusev, 2001; Swadlow et al., 2002) and/or the activation of T-type Ca^{2+} channels that have a lower threshold than Na^+ channels (see section 10.3 below). Importantly, the augmenting influence that interspike interval can have on retinal efficacy does not depend on the overall strength of connection between retinal and lateral geniculate cells (Usrey et al., 1998), since cell pairs that are strongly connected (high

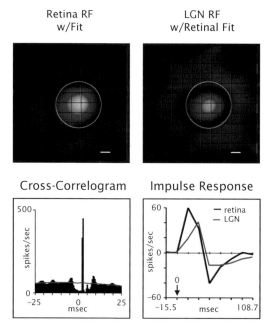

Retina RF
w/Fit

LGN RF
w/Retinal Fit

Cross-Correlogram

Impulse Response

FIGURE 10.1 Receptive fields, impulse responses, and cross-correlogram of a pair of monosynaptically connected neurons in the retina and lateral geniculate nucleus of the cat. White-noise receptive field maps (see Reid et al., 1997 for method) show the receptive fields of a retinal ganglion cell (top left) and geniculate cell (top right) recorded simultaneously. On regions indicated in red, off in blue. The thin circle in both panels corresponds to a fit of the center of the retinal ganglion cell's receptive field (1.75 σ, or standard deviations, from the peak of the best fitting Gaussian). The cross-correlogram (lower left) shows the relative activity of the two cells. Retinal spikes occur at time zero and the short-latency peak to the right of zero indicates that many retinal spikes trigger a spike in the geniculate cell with a latency of ~2.5 msec. This peak provides evidence that the retinal ganglion cell and geniculate cell are monosynaptically connected. The impulse response (lower right) shows the time-course of visual response for the retinal ganglion cell (black) and geniculate cell (red). The peak of the visual response of the geniculate cell is ~15.5 msec slower than the peak response of the retinal ganglion cell.

Abbreviations: RF: receptive field; *LGN:* lateral geniculate nucleus. Modified from Usrey et al., 1999.

probability of spike transfer) display the same degree of interspike interval dependent enhancement as do cell pairs that are weakly connected.

The relationship between retinal interspike interval and spike efficacy has been demonstrated in both anesthetized (Mastronarde, 1987; Usrey et al., 1998; Levine and Cleland, 2001; Rowe and Fischer, 2001; Carandini et al., 2007) and alert animals (Weyand, 2007). The dynamics of retinogeniculate transmission have also been studied in vitro with brain slices that include the lateral geniculate nucleus and cut retinal axons (Chen et al., 2002; Lo et al., 2002). The in vitro approach has the advantage that the mechanism(s) underlying the dynamic properties of synaptic transmission can be studied. Using the in vitro approach, Chen et al. (2002) demonstrated that retinogeniculate synapses undergo paired-pulse depression (see Chapter 4). Although the two sets of findings—(1) short interspike intervals increase the likelihood of successful retinogeniculate communication, and (2) short interspike intervals lead to synaptic depression—would seem to be at odds with each other, it is worth noting that retinal firing rates measured in vivo are typically sufficiently high as to place retinogeniculate

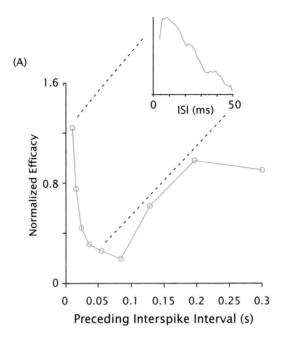

(A)

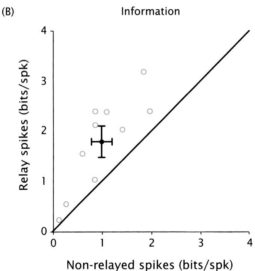

(B)

FIGURE 10.2 Relationship between cat retinal interspike interval on evoking geniculate action potentials relayed to cortex and the relative information of relayed spikes versus non-relayed spikes. A. Plot showing the average efficacy of two retinal spikes as they occur with increasing interspike interval. Efficacy is equal to the percentage of retinal spikes that elicit a geniculate spike. Second retinal spikes are much more effective than first retinal spikes at very short interspike intervals. As the interspike interval increases, the efficacy of second retinal spikes decreases until ~30 msec when second spike efficacy is approximately the same as first spike efficacy. At longer interspike intervals, efficacy increases. Presumably due to release of synaptic depression. B. Scatter plot comparing the amount of information (bits/spike) contained in relayed and non-relayed retinal spikes. Relayed spikes carry significantly greater information than non-relayed spikes. Cross-hairs indicate mean and SEM. A. Modified from Usrey et al., 1998; Alitto et al., 2019. B. Modified from Rathbun et al., 2010.

synapses in a state of sustained depression. In this state, postsynaptic temporal summation can fully account for the measured relationship between retinal interspike interval and efficacy in evoking lateral geniculate spikes (Carandini et al., 2007). As mentioned above, a release from depression likely contributes to the second phase of enhanced efficacy that occurs when retinal spikes follow longer interspike intervals (>80 msec). Although longer interspike intervals are much less common than short interspike intervals, specific patterns of spatiotemporal visual stimulation, as occur in natural scenes, can increase their likelihood.

The identified relationship between retinal interspike interval and retinal efficacy in evoking spikes in the lateral geniculate nucleus leads to the prediction that lateral geniculate responses to repeated patterns of visual stimulation should be reliable and consistent. As predicted, Kara et al. (2000) found that lateral geniculate responses are much less variable than that of a Poisson process. Similar results have been reported for recordings from lateral geniculate and hippocampal slices using short segments of natural stimulus trains (Dobrunz and Stevens, 1999; Chen et al., 2002). Moreover, comparisons demonstrate that retinal spikes that are relayed to cortex carry more information than retinal spikes not relayed to cortex (Figure 10.2B; Rathbun et al., 2010; Wang et al., 2010). Taken together, these results provide support for the idea that the dynamics of retinogeniculate interactions are consistent and reliable. Of course, behavioral state and changes in the activity levels of modulatory inputs to lateral geniculate neurons are likely to adjust the gain of lateral geniculate responses to retinal input.

10.2 Retinal Divergence and Lateral Geniculate Synchrony

Individual retinal ganglion cells innervate more than one lateral geniculate neuron, and individual lateral geniculate neurons typically receive input from more than one retinal ganglion cell. Thus, the pathway from retina to the lateral geniculate nucleus is both divergent and convergent. While some lateral geniculate neurons in the cat receive all of their retinal input from just one retinal ganglion cell (Mastronarde, 1987), most lateral geniculate neurons are thought to receive convergent input from a small number of ganglion cells (3–5), with partially overlapping receptive fields of the same on- or off-center and X or Y types (Levick et al., 1972; Mastronarde, 1987; Usrey et al., 1999). How individual lateral geniculate neurons integrate these convergent inputs is an important question to answer and one that deserves future attention. More is known about the effects of divergent connections (reviewed in Usrey and Reid, 1999; Usrey, 2002a; Usrey and Alitto, 2015). Based on paired recordings from the retina and lateral geniculate nucleus, Cleland (1986) proposed the idea that retinal ganglion cells with divergent axons should induce synchronous responses among the postsynaptic lateral geniculate neurons. Synchrony was later found in the cat with multielectrode recordings of lateral geniculate neurons (Alonso et al., 1996) and demonstrated to be the result of common retinal input (Figure 10.3; Usrey et al., 1998). Synchrony resulting from anatomical divergence in the lateral geniculate nucleus is strong and tight, with up to 30% of the spikes

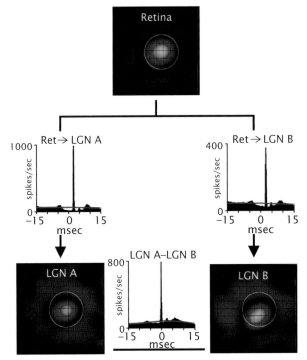

FIGURE 10.3 Receptive fields and cross-correlograms of two geniculate neurons (cell A and cell B) in the cat that receive common input from a retinal ganglion cell (all cells recorded from simultaneously). Geniculate neurons that receive common retinal input fire many synchronous spikes. Receptive fields were mapped using a white noise stimulus. On responses shown in red, off responses in blue. The circle over all receptive fields corresponds to a Gaussian fit of the center of the retinal cell's receptive field. The correlograms at the sides of the figure have short latency (~2.5 msec) peaks indicating that the retinal ganglion cell provided monosynaptic input to both geniculate neurons. The peak in the correlogram at the bottom of the figure shows that the two geniculate cells fire many spikes simultaneously (<1 msec). Red traces superimposed on the correlograms correspond to the stimulus-dependent firing of the cells ("shuffle" correlogram; a correlogram made after shifting the activity of one cell by the amount of time corresponding to the stimulus cycle).

Abbreviations: Ret: retinal ganglion cell; *LGN:* lateral geniculate nucleus. Modified from Usrey et al., 1998.

from two lateral geniculate cells with common input occurring within less than 1 msec of each other. With these features in mind, it is noteworthy that divergence is present in both the X and Y cell pathways from the retina to the lateral geniculate nucleus, with individual retinal X axons contacting approximately 4 lateral geniculate relay cells compared to individual Y axons contacting upward of 20–30 cells (Friedlander et al., 1981). As a consequence, it seems likely that size of the ensembles of synchronously active Y cells in the lateral geniculate nucleus is greater than that of X cells, a difference that may serve to strengthen the Y pathway to cortex.

Synchrony between lateral geniculate neurons is also affected by retinal interspike interval. When retinal spikes occur with interspike intervals of less than 30 msec, the second spike of the pair can be up to 12 times more effective than the first spike in driving synchronous responses (Usrey et al., 1998). Because of this, when the firing rate of an individual retinal ganglion cell increases rapidly, as occurs during visual stimulation, there is a dramatic increase in synchronous lateral geniculate spikes. Thus, there is a partial transformation in information coding from a single-cell rate code in the retina to a population temporal code in the lateral geniculate nucleus (Dan et al., 1998). Whether or not these synchronous lateral geniculate spikes play a role in visual processing depends on whether or not there is a cortical mechanism for preferentially detecting coincident spikes (see section 10.4.1 below).

Studies investigating burst activity in the thalamus have typically relied on recordings from single cells, one at a time. However, modulatory inputs, because of their divergent inputs that simultaneously modulate numerous relay cells, might be capable of priming burst activity allowing for the generation of a synchronous burst across many relay cells in response to a sensory stimulus. As described above and given the divergence of retinal input to geniculate cells, the convergence of these cells onto cortical neurons and the ability of bursts to strongly activate cortex would combine to create an especially strong activation of cortex.

10.3 Corticothalamic Feedback Influences on Lateral Geniculate Activity

As described in previous chapters, lateral geniculate neurons receive nonretinal input from a variety of sources including the thalamic reticular nucleus, various regions of the brainstem, and, an emphasis of this section, layer 6 of visual cortex. Despite the prominence of layer 6 corticogeniculate input—feedback axons provide approximately 5–10 times more synapses onto lateral geniculate neurons than do retinal axons (Guillery, 1969a; Erisir et al., 1997a,b; Van Horn et al., 2000)—we lack a firm understanding of what role corticogeniculate feedback plays for vision. This lack of knowledge likely reflects, at least in part, limitations associated with prior approaches for studying corticogeniculate feedback; most studies were conducted in anesthetized animals with impaired cortical activity, and most of these utilized large-scale and nonreversible means of cortical inactivation or activation that obscured details of topography or functional heterogeneity among different classes of thalamocortical neurons. Since lateral geniculate receptive fields bear little relation to those of their layer 6 input, cortical feedback likely plays a modulatory rather than driving role in its effect on lateral geniculate activity, perhaps serving to influence the temporal properties and/or gain of lateral geniculate responses rather than the structure of lateral geniculate receptive fields. Among the proposed functions of corticogeniculate feedback are: (1) adjusting the gain or timing of lateral geniculate responses to retinal input; (2) shifting lateral geniculate neurons between burst and tonic modes of firing (Sherman, 1996, 2001; Kirchgessner et al., 2020); and (3) increasing the correlated activity of ensembles of lateral geniculate neurons (Sillito et al., 1994; see also Weliky, 1999; Jones, 2001). Importantly, as feedback has the opportunity to affect relay cell activity via monosynaptic excitation and/or disynaptic inhibition, feedback effects may vary depending on synaptic dynamics that vary with firing rate (Crandall et al., 2015; Kirchgessner et al., 2020) and/or the spatial relationship between the receptive fields of feedback cells and their lateral geniculate target cells.

There is experimental support for the first proposed role—feedback serves to adjust the gain or timing of lateral geniculate responses to retinal input (Schmielau and Singer, 1977; Tsumoto et al., 1978; Molotchnikoff et al., 1984; Gulyás et al., 1990; McClurkin et al., 1994; Funke et al., 1996; Rao and Ballard, 1999; Przybyszewski et al., 2000; Olsen et al., 2012; Hasse and Briggs, 2017b; Murphy et al., 2020); however, most of this support comes from experiments in which feedback is manipulated in a global fashion involving large areas of cortex (i.e., with cooling, aspiration, pharmacology, or optogenetics) in anesthetized animals, a less than ideal approach for studying the function of feedback projections that may have spatial (e.g., center/surround) topography (Wang et al., 2018; Sanchez et al., 2021; see Figure 7.1B), functionally distinct parallel streams (Briggs and Usrey, 2009), and depend on brain state. For example, Figure 7.1B indicates a simple arrangement whereby the layer 6 feedback provides monosynaptic excitation and disynaptic inhibition to all relay cells, whereas Figure 7.1C shows a different pattern. However, global excitation would have the same effect on relay cells—monosynaptic excitation and disynaptic inhibition—for both patterns shown in Figure 7.1B,C, thereby obscuring the spatial properties of the feedback. Likewise, global suppression would equally obscure the differences between Figure 7.1B and C.

It is worth emphasizing that the influence of feedback may be excitatory or suppressive, because feedback axons terminate monosynaptically onto relay cells as well as onto local interneurons and neurons in the thalamic reticular nucleus which, in turn, make GABAergic synapse with thalamic relay cells (Figure 7.1B,C). Indeed, experimental evidence supports the notion of feedback providing both an excitatory and suppressive influence as dictated by the spatial alignment of feedback projections, whereby feedback neurons with receptive fields that are in topographic alignment with those of a geniculate neuron provide net excitation to the geniculate neuron and other feedback neurons with receptive fields located further away provide net suppression (Jones et al., 2012; Wang et al., 2018; Sanchez et al., 2021).

The second proposed role—feedback influences the burst versus tonic activity mode of lateral geniculate neurons—is based on the time/voltage (hyperpolarization) relationship that governs the de-inactivation of T-type Ca^{2+} channels (Jahnsen and Llinás, 1984a,b; Lo et al., 1991; Huguenard and McCormick, 1992; McCormick and Huguenard, 1992). As discussed in greater detail in Chapter 3, the low-threshold spike that underlies bursts requires de-inactivation of T-type Ca^{2+} channels. If the T-type Ca^{2+} channels are de-inactivated, then subsequent suprathreshold depolarization results in a Ca^{2+} spike upon which rides a short train of Na^+–based actions potentials, and this is a burst. In contrast, if the T-type Ca^{2+} channels are inactivated, then the same depolarization results in tonic spiking. Corticogeniculate feedback is thought to affect the resting potential of thalamic neurons and, consequently, shifts thalamic neurons between burst and tonic activity modes (Sherman, 1996, 2001), which affects both the nature of the signals sent to cortex and the likelihood of evoking cortical responses (see section 10.4.2).

The third proposed role—feedback increases correlated activity of ensembles of lateral geniculate neurons—has been championed by Sillito and colleagues (Sillito et al., 1994). By recording from pairs of lateral geniculate cells with nearby, but offset, receptive fields, they found that the correlated activity between the cells was both increased and sharpened during visual stimulation when the corticogeniculate feedback pathway was intact compared to when it was inactivated. Subsequent work has suggested an alternative explanation for these correlations based on slow covariations in the resting potential between geniculate cells (Brody, 1998); thus, further work is needed to rule out this possibility.

10.4 Significance of Different Patterns of Cortical and Thalamic Activity

With the increasing use of multielectrode recording techniques to study synaptically connected neurons in the brain, researchers have the tools needed to answer questions about how the thalamus communicates with the cerebral cortex in vivo. These techniques have been applied successfully to study thalamocortical connections in vivo in the visual (Tanaka, 1985; Reid and Alonso, 1995; Alonso et al., 1996, 2001; Usrey et al., 2000; Sedigh-Sarvestani et al., 2017), somatosensory (Roy and Alloway, 2001; Swadlow and Gusev, 2001; Swadlow et al., 2002; Bruno and Sakmann, 2006), and auditory systems (Miller et al., 2001a,b). Paired thalamocortical recordings can be used to answer questions about the specificity and strength of thalamic connections as well as the role of spike timing in thalamocortical communication.

10.4.1 Thalamic Interspike Interval and Thalamocortical Communication

Simultaneous recordings from monosynaptically connected neurons in the lateral geniculate nucleus and visual cortex have also been used to examine how lateral geniculate spike timing affects cortical responses. Similar to spikes in the retinogeniculate pathway, the preceding interspike interval of lateral geniculate spikes has a significant influence on whether or not they are successful in evoking a cortical spike. Specifically, the second spike of a pair has an increased probability of driving a cortical spike when it arrives following a relatively short interspike interval (Usrey et al., 2000). This enhanced probability is greatest at the shortest interspike intervals measured and decreases with interspike intervals up to approximately 15 msec. Similar to the retinogeniculate pathway, in vitro studies report that the geniculocortical pathway to layer 4 undergoes paired-pulse depression (Stratford et al., 1996; Boudreau and Ferster, 2005; see also Gil et al., 1999; Chung et al., 2002; Lee and Sherman, 2008; Viaene et al., 2011b), and like the retinogeniculate pathway, this depression appears to be saturated with firing rates typical for visual stimulation (Boudreau and Ferster, 2005). Still, lateral geniculate firing patterns do include periods of inactivity where synaptic depression could modulate the strength of geniculocortical communication. Although technically challenging, paired lateral geniculate and cortical whole-cell recordings provide a more complete understanding of the role of thalamic spike timing and synaptic depression on cortical responses (Hirsch et al., 1995; Ferster et al., 1996; Azouz and Gray, 2000; Chung et al., 2002; Sedigh-Sarvestani et al., 2017).

Cortical layer 4 neurons receive convergent input from several lateral geniculate neurons. The degree of convergence varies between species, but in the cat estimates suggest that individual layer 4 neurons receive convergent input from approximately 30 lateral geniculate neurons (reviewed in Peters and Payne, 1993; Reid et al., 2001). By recording simultaneously from two lateral geniculate cells that provide synaptic input to a common cortical cell, researchers have examined how cortical neurons respond to spikes arriving from convergent inputs (Alonso et al., 1996; Usrey et al., 2000; Figure 10.4). Similar studies have been performed in the somatosensory system (Roy and Alloway, 2001). In both systems, convergent inputs interact in a reinforcing fashion over very brief windows of time to drive cortical spikes. Reinforcement is maximal for spikes that arrive within 1 msec of each other and then decrease until approximately 7 msec (~2.5 msec time constant) when spikes from two separate thalamic cells appear independent of each other. As mentioned above,

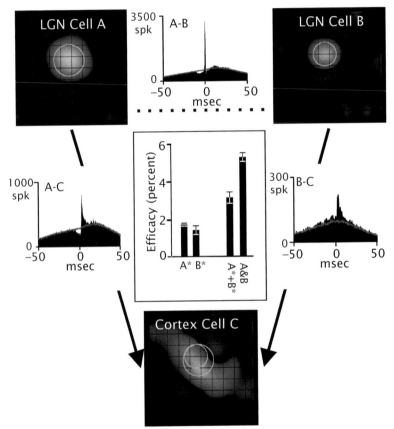

FIGURE 10.4 Receptive fields and cross-correlograms of two geniculate neurons (cell A and cell B) that provide convergent input to a common cortical neuron (cell C; all cells recorded from simultaneously). Geniculate inputs that arrive simultaneously to a cortical neuron interact synergistically to drive a cortical response. Receptive fields were mapped using a white noise stimulus. On responses shown in red, off responses in blue. The circle over the cortical cell's receptive field corresponds to Gaussian fits of the centers of the geniculate neurons' receptive fields. The correlograms at the sides of the figure show that both geniculate neurons provide monosynaptic input to the cortical neuron. Red traces superimposed on the correlograms correspond to the stimulus-dependent firing of the cells. Inset. To examine the interactions of converging inputs, geniculate spikes were divided into three groups: geniculate spikes that occurred within 1 msec of each other (A & B), geniculate spikes that occurred in cell A, but not cell B (A*), and geniculate spikes that occurred in cell B, but not cell A (B*). The efficacy (% of spikes that drive a cortical spike) of simultaneous geniculate spikes (A & B) is 70% greater than that expected if geniculate spikes interact in a linear fashion (A* + B*).

Abbreviation: *LGN*: lateral geniculate nucleus. Modified from Alonso et al., 1996.

retinogeniculate divergence underlies tight spike synchrony (< 1 msec) between lateral geniculate cells (Alonso et al., 1996; Usrey et al., 1998). This result, taken together with results
from geniculocortical recordings, suggests that layer 4 neurons have the means to respond
selectively to these synchronous spikes. Although there has been an ongoing debate about
whether or not precise timing of presynaptic inputs is important for sensory processing in
the cerebral cortex (Shadlen and Movshon, 1999), results from the experiments described
above indicate that spike timing is important for thalamocortical connections.

10.4.2 Thalamic Bursts and Thalamocortical Communication

As mentioned above in section 10.3, thalamic neurons have two response modes—burst
and tonic—that depend on the state of the T-type Ca^{2+} channels: burst mode prevails when
T channels are in the de-inactivated state and tonic mode prevails when the channels are
in the inactivated state. Depending on the state of these channels, thalamic neurons will
therefore produce very different responses to the same sensory stimulus (Figure 10.5; also
see Figure 3.7). In burst mode, spontaneous activity is low and sensory stimulation evokes

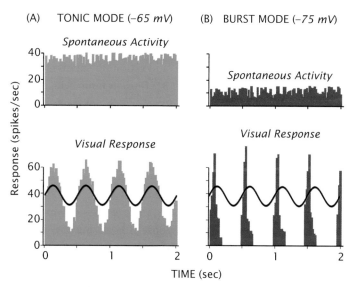

FIGURE 10.5 Average response histograms of a representative relay cell in the lateral geniculate nucleus
of a lightly anesthetized cat to a sinusoidal grating drifted through the cell's receptive field. The black
traces on the bottom histograms reflect the sinusoidal changes in luminous contrast with time. The top
histograms represent spontaneous activity in the absence of a grating whereby the stimulus is a feature-
less background of the same average luminance as the grating, and the bottom histograms represent responses to the drifting grating. Current was injected into the cell through the recording electrode to alter
the membrane potential. A. Tonic firing mode. The current injection was adjusted so that the membrane
potential without visual stimulation averaged –65 mV, promoting tonic firing, because I_T is mostly inactivated at this membrane potential. B. Burst firing mode. The current injection was adjusted to the more
hyperpolarized level of –75 mV, permitting de-inactivation of I_T and promoting burst firing. Note that the
response profile during the visual response in tonic mode looks like a sine wave, but the companion response during burst mode does not. Note also that the spontaneous activity is higher during tonic than
during burst firing. Note finally that the response latency is reduced during burst firing, indicating a phase
advance in the response relative to tonic firing. Redrawn from Sherman and Guillery, 2013.

a transient volley of spikes that are invariant to stimulus strength. In contrast, spontaneous activity is higher when cells are in tonic mode and sensory stimulation evokes a train of spikes that vary in frequency with the strength of the stimulus. Because bursts occur more frequently when animals are drowsy or inattentive (Guido and Weyand, 1995), the sudden occurrence of a stimulus-evoked burst (Lesica and Stanley, 2004; Alitto et al., 2005; Denning and Reinagel, 2005) has been proposed to serve as a "wake-up call" for the cortex (Sherman, 1996, 2001), a call that engages cortical circuits for the subsequent arrival of tonic spikes that carry higher fidelity information.

A key aspect of the "wake-up call" hypothesis is that the cortex responds particularly well to the arrival of burst spikes, a prediction that is supported by results demonstrating the cortex responds best to thalamic spikes that arrive with the shortest of interspike intervals (Usrey et al., 2000). In addition, because of the time–voltage relationship that is necessary to de-inactivate T-type Ca^{2+} channels, bursts can only occur after a period of hyperpolarization to de-inactivate the underlying Ca^{2+} channels; such a hyperpolarized period means no firing of action potentials, and this silent period means that thalamo-cortical synapses are relieved of their depression. Thus, the first spike in a burst activates a synapse that will evoke its largest EPSP, and the high-frequency firing allows for the temporal summation of thalamocortical inputs. Together, these processes result in max-imum cortical activation, consistent with experimental results (Swadlow and Gusev, 2001; Swadlow et al., 2002).

During tonic firing, relay cell Na^+ spikes are directly linked to an EPSP in that cell, but, in burst mode, the link between the EPSP and action potentials is indirect via the all-or-none low-threshold Ca^{2+} spike (Figure 3.7A,B). Thus, in tonic firing, larger EPSPs evoke higher firing rates; however, larger EPSPs in burst mode do not evoke larger low-threshold spikes and thus do not evoke higher firing rates. Thus, the input–output relationship for tonic firing is fairly linear, whereas that for burst firing is nonlinear, appearing like a step function (Figure 3.7E).

Studies of geniculate responses to visual stimuli in vivo indicate that both response modes transmit roughly equal levels of information (Reinagel et al., 1999). Nonetheless, the quality of the information differs between modes, three aspects of which are illustrated by Figure 10.5. First, during both modes, the response is brisk, but the response profile to a sinusoidal input is more sinusoidal during tonic than during burst firing (Figure 10.5A,B, *bottom*), thus reflecting better linear summation (see also Figure 3.7E). Second, spontaneous activity is higher during tonic firing (Figure 10.5A,B *top*), which helps maintain linearity by minimizing rectification of the response. More to the point, spontaneous activity represents noise against which the visual response (i.e., the signal) must be detected. The histograms of Figure 10.5 suggest that the signal-to-noise ratio and thus detectability is higher during burst than during tonic firing. By using techniques of signal detection theory (Green and Swets, 1966; Macmillan and Creelman, 1991), it has been shown that burst responses provide en-hanced stimulus detectability (Guido et al., 1995). Thus tonic firing affords better linearity, whereas burst firing supports better signal detection, and other differences in information relayed between firing modes have also been reported (Mukherjee and Kaplan, 1995; Smith et al., 2000). Third, burst firing provides a phase advance in the response.

Tonic firing minimizes nonlinear distortions in the relay, thereby supporting a more faithful reconstruction of the visual world. Burst mode maximizes initial stimulus detection, perhaps as a sort of "wake-up call" that something has changed in the environment (Sherman, 1996, 2001). It may seem odd to think of a burst as a "wake-up call" when, during slow wave sleep, bursting dominates the responses of thalamic relay cells. Perhaps this notion is wrong, but it seems at least equally plausible that the rhythmic, synchronized bursting of relay cells during sleep reflects a very different message for cortex, one that indicates a complete lack of information being relayed from the periphery. This notion of a "wake-up call" is much like the earlier "searchlight hypothesis" of Crick (1984). He suggested that neurons in the thalamic reticular nucleus powerfully inhibit relay cells, thereby de-inactivating I_T in these cells and causing the next excitatory input to activate a burst. This is similar to the notion proposed here, although Crick includes more specific suggestions for control of the reticular searchlight and the effect of bursting on dynamic cell assemblies (see Crick, 1984, for details).

For the current proposal, one example to consider is a geniculate cell's response to the sudden appearance of a novel object in a previously uninteresting or unattended region of visual space. If the cell were in burst mode, it would better signal the detection of this object and still permit a crude, initial analysis of it. Once the change is detected, the neuron could be switched to tonic mode so that the new object could be analyzed more faithfully. It is also notable, as described above, that bursts more effectively activate EPSPs in their target cortical cells.

10.4.3 Bursting in Layer 5 Corticofugal Cells

10.4.3.1 Basic Properties of Bursting in Layer 5 Corticofugal Cells

In Chapter 3 section 3.4.2, we pointed out that tonic and burst firing also exists in layer 5 corticofugal cells that in many ways resembles these different firing modes in thalamic relay cells, although different Ca^{2+} channels are involved (Larkum et al., 1999, 2007; Llano and Sherman, 2009; Suzuki and Larkum, 2020). One similarity to thalamic bursting is that these layer 5 cells must be suitably hyperpolarized for a period of time to de-inactivate the Ca^{2+} channels so that a burst can be evoked, and, during such hyperpolarization, these cells would not fire action potentials. We can infer several implications from this, although we emphasize that these speculations have not yet been experimentally verified. First, because the layer 5 inputs to higher order thalamic relay cells have driver characteristics (Sherman and Guillery, 2013; Sherman, 2016), their corticothalamic synapses show paired-pulse depression (see Chapter 4). Second, the requisite period of hyperpolarization and lack of firing would relieve such depression and mean that a burst would activate the corticothalamic synapses with maximum amplitude EPSPs and temporal summation. Thus, just as thalamocortical bursts maximally activate cortex, we predict that these layer 5 bursts maximally activate their target thalamic cells.

Perhaps this acts as a "wake-up call" for the transthalamic pathway much as it does for thalamocortical processing, as suggested above. That is, when the layer 5 cells are quiescent due to hyperpolarization and little or no information is being passed through the transthalamic circuit, a sufficiently large depolarization indicating a change in afferent information will activate a burst in these cells, and this would maximally activate the transthalamic pathway.

There are several problems and unresolved issues with this speculation. One is that bursting in these cells, as in thalamic relay cells, likely involves nonlinear distortion in the message transmitted. Thus, if indeed bursting of these layer 5 cells evokes a "wake-up call" for the transthalamic pathway, it would seem useful to then switch firing to tonic mode for more faithful transmission of the messages. Second, because these layer 5 projections involve branching to many extrathalamic motor sites as well as thalamus, what is the implication of this strong activation of these sites by the bursts? Is this a way to "jump start" the initial phase of a motor action? Clearly, there is a great deal we need to learn about the properties of these layer 5 corticofugal cells, especially since they are the sole means by which cortex can plausibly influence behavior (Sherman and Usrey, 2021).

10.4.3.2 Control of Bursting in These Layer 5 Cells

We speculated above how the burst/tonic transition might be controlled (e.g., via layer 6 feedback circuitry), and the question now is how bursting is controlled in these layer 5 cells. There are several pieces of evidence, largely from the work of Matthew Larkum and his colleagues (Larkum et al., 1999, 2004; Suzuki and Larkum, 2020) that suggest possible control mechanisms. Bursting in these cells occurs when there is conjoint synaptic activation of their apical tufts in layer 1 and input to more proximal dendrites. The summed resultant depolarization, which typically involves a back-propagating action potential, is sufficient to activate the Ca^{2+} spike in the apical dendrite, thereby producing a burst. This requires coupling between the apical dendritic tufts in layer 1 and the main shaft of the apical dendrite. This coupling can be broken by anesthetics but, more interestingly, also by blockers of metabotropic receptors for either acetylcholine (i.e., muscarinic receptors) or glutamate (i.e., metabotropic glutamate receptors) (Suzuki and Larkum, 2020). It thus follows that cholinergic input from the basal forebrain, which corresponds mostly to overall behavioral state, or that from glutamatergic modulators, which can provide more specific and topographic control of activity in these layer 5 corticofugal cells, can provide necessary conditions for bursting in these cells.

It has been suggested that the main source of input to the layer 1 tufts of the layer 5 cells in question is a feedback projection from other cortical areas (Suzuki and Larkum, 2020) and that the control of these layer 5 cells is all about the extent to which these feedback circuits control the main output of a cortical column. This would depend heavily on the dendritic tufts in layer 1 of the layer 5 output cells to be electrotonically coupled to the main dendritic shafts, coupling that can be provided by activation of metabotropic glutamate receptors in these cells. It then is of some interest to identify the sources of input of glutamatergic modulators to these cells that can activate metabotropic glutamate receptors.

One source to consider is input from other cortical areas. However, limited data from connections between auditory and visual cortical areas in both directions indicate that such inputs to layer 5b, where the corticofugal cells reside, are overwhelmingly driver, and these inputs do not activate metabotropic glutamate receptors (Covic and Sherman, 2011;

DePasquale and Sherman, 2011). However, these experiments, which recorded EPSPs in cell bodies, likely failed to detect inputs onto distal tufts of these cells in layer 1, and so it remains possible that these inputs to layer 1 include glutamatergic modulators that activate metabotropic glutamate receptors.

Another source is local glutamatergic inputs, and to date there are no data we know of that addresses this possibility.

A final source of input is from thalamus, and, for the most part, relevant data are lacking, but there are some. Feedforward thalamocortical projections so far studied to layer 5 from first order sensory thalamic nuclei to primary sensory cortices provide only driver input to these cells (Viaene et al., 2011b). Feedforward projections from higher order thalamic nuclei to layer 5 cells in higher order cortical areas have to date not been described. However, the feedback projections from higher order thalamic nuclei to first order cortical areas do provide modulatory inputs to these layer 5 cells that activate the requisite metabotropic glutamate receptors (Viaene et al., 2011a). It is also noteworthy that bursts in these layer 5 cells could be evoked via co-activation of synaptic inputs to proximal dendrites *and* inputs to the distal dendritic processes in layer 1, and these latter inputs are thought to be feedback projections from other cortical areas (Suzuki and Larkum, 2020). One purely speculative possibility to consider is that bursting in these layer 5 cells is a reflection of feedback corticocortical circuitry involving both direct projections to layer 1 and transthalamic inputs that provide activation of metabotropic glutamate receptors on the output layer 5 cells.

10.5 Summary and Concluding Remarks

Spike timing plays a prominent role in thalamic processing and thalamocortical interactions. Individual EPSPs from afferent inputs to thalamic neurons are generally too weak to evoke suprathreshold responses on their own. However, EPSPs that arrive within approximately 30 msec of a prior EPSP have a much greater probability of evoking postsynaptic spikes, presumably because of temporal summation. As a consequence, thalamic neurons produce fewer spikes than their afferent inputs, and there is an improvement in the signal-to-noise ratio between the spike trains of pre- and postsynaptic neurons. In the visual system, branching axons from the retina deliver spikes simultaneously to multiple neurons in the lateral geniculate nucleus, and these simultaneously arriving spikes produce synchronous spikes among their ensemble targets. These synchronous spikes are particularly effective in evoking postsynaptic spikes in the cortex. Also, as with afferent inputs to thalamic neurons, thalamic inputs to cortical neurons summate over relatively short interspike intervals to enhance thalamocortical communication.

A ubiquitous property of thalamic neurons is the ability to produce tonic spikes or bursts of spikes depending on the neuron's membrane potential history. The underlying mechanism for the generation of burst spikes is the hyperpolarization-dependent de-inactivation of T-type Ca^{2+} channels. Once de-inactivated, these channels are poised to respond to subsequent EPSPs by generating a low-threshold Ca^{2+} spike which can then trigger a burst of Na^+-dependent spikes. Several factors can affect the de-inactivation state of the T-type Ca^{2+}

channels, including specific spatiotemporal patterns of sensory stimulation, corticothalamic feedback, and brain state. Because thalamic bursts can encode sensory information and they are particularly effective in driving cortical responses, bursts have been argued to serve as a wake-up call for the cortex, alerting the cortex to the arrival of sensory information that generally follows in the form of tonic spiking.

10.6 Some Outstanding Questions

1. What is the functional significance of bursting in thalamic relay cells and layer 5 cortico-fugal cells? We have offered speculation, but the question remains open.
2. We have argued that the divergence of retinogeniculate inputs and convergence of their target geniculocortical projections boosts retino-geniculo-cortical functioning. Is this accurate, and, if so, is it a general feature of thalamocortical circuitry?
3. What are the patterns of layer 6 corticothalamic circuitry seen at the single cell level? Are there numerous patterns as depicted in Figure 7.1B,C, and, if so, how do these correlate with the parallel feedback patterns?
4. Feedback projections utilize ionotropic and metabotropic receptors to influence postsynaptic neurons. What is the functional significance of having these two classes of receptors, and how do they interact to affect relay cell activity?
5. How common are synchronized bursts evoked by sensory stimulation seen in thalamic relay cells?

11

Parallel Processing of Sensory Signals to Cortex

The notion of parallel pathways from the periphery into the brain is not new. In the early 19th century, Charles Bell and François Magendie separately argued for what Johannes Müller later termed the "specific energy of nerves" (see Clarke and O'Malley, (1968)). This term was a precursor to the idea that individual sensory axons are "labeled lines" and the notion of parallel pathways for providing distinct signals to the brain. This chapter includes and builds on previous discussions of parallel processing as it relates to the thalamus and thalamocortical circuits (Briggs and Usrey, 2011; Usrey and Alitto, 2015; Usrey and Sherman, 2019; Sanchez et al., 2020). Because much of what we know about parallel processing has been gained from studies of the visual system, we will focus on the visual system as a model, emphasizing the organization of parallel processing streams in the feline, primate, and rodent. With this knowledge, we then speculate on questions of evolution, homology, and parallel pathways interconnecting thalamus and cortex.

11.1 Felines

11.1.1 Distinct Classes of Retinal Ganglion Cells Establish Parallel Pathways in the Cat

Stephen Kuffler (1952, 1953) reported that retinal ganglion cells in the cat have receptive fields with a stereotypical center-surround organization. He described two types of cells, "on cells" that are maximally excited by small spots of light surrounded by darkness and "off cells" that are maximally excited by small spots of darkness surround by light. Importantly, on and off cells were found at every position across the entire retina, suggesting that there are at least two complete, but different, representations of the visual world that travel together (i.e., "in parallel") in the axons of each optic nerve. The notion of two complete but different representations rapidly grew as several additional functional classes of ganglion cells were identified, each with distinct response properties and each tiling the retina. For instance, Enroth-Cugell and Robson (1966) demonstrated that, within both the on and off cell classes, there are two additional cell categories: X and Y. The visual responses of X cells could be accounted for by simple summation of their excitatory and inhibitory inputs (linear summation), whereas the responses of Y cells indicated more complex interactions between inputs (nonlinear summation). Although this distinction may seem somewhat abstract, X and Y cells were later found to differ in several additional and important ways: X cells (called *beta cells* by the anatomists) have smaller dendritic fields, smaller receptive fields, longer visual response latencies, slower axonal conduction velocities, respond in a sustained fashion to stationary stimuli, and prefer stimuli with lower temporal frequencies, whereas Y cells (called *alpha cells* by the anatomists) have larger dendritic fields, larger receptive fields, shorter response latencies, faster axonal conduction velocities, respond transiently to stationary stimuli, and follow stimuli at higher temporal frequencies (reviewed in Lennie, 1980; Sherman and Guillery, 2006; Usrey and Alitto, 2015). Many of these cell-class specific distinctions are also evident in other sensory modalities. For instance, differences in receptive field size, sustained versus transient responses, and stimulus temporal frequency (flutter vs. vibration) are also used to identify and distinguish tactile receptors (e.g., Pacinian corpuscles, Meissner corpuscles, Ruffini endings, and Merkle discs) that supply the medial lemniscal pathway to the ventral posterior nucleus of the thalamus.

Distinct from the X and Y cells of the retina, a third group of cells, W cells, broadly include all of the cell types that do not fit neatly within the X and Y cell categories. We now know that W cells, as a group, are comprised of several unique cell classes. Although each of these cell classes is believed to form a complete mosaic across the retina, we know much less about the multitude of W cells compared to X and Y cells, except that some have on/off responses and others have direction selectivity, among other properties.

11.1.2 Parallel Processing in the Cat Lateral Geniculate Nucleus

Similar to the retina, the cat lateral geniculate nucleus contains X cells, Y cells, and W cells. X cells have highly linear center–surround receptive fields that are nonselective for either stimulus orientation or direction of motion (Stone, 1983). Although X cells in the geniculate are generally regarded as having linear responses, they are not completely linear as they often display a degree of contrast gain control and extraclassical suppression (Jones et al., 2000; Alitto and Usrey, 2004). Still, Y cells exhibit stronger nonlinear responses, greater contrast gain control, and more robust extraclassical suppression (Rathbun et al., 2016; Fisher et al., 2017). Two main features distinguish X and Y receptive fields: both have relatively linear center–surround organization, but those of Y cells are roughly three times larger in diameter when adjusted for eccentricity (Sherman and Spear, 1982; Sherman, 1985), and the main nonlinearity of Y cells is in the form of nonlinear subunits distributed throughout the receptive field, subunits not seen in X receptive fields (Hochstein and Shapley, 1976). Similar to their sources of input from the retina, geniculate Y cells compared to X cells have larger receptive fields, shorter visual response latencies, faster conducting axons, more transient responses to visual stimuli, and can follow stimuli at higher temporal frequencies (Cleland et al., 1971b; Fukada, 1971; Hoffmann et al., 1972; Ikeda and Wright, 1972; Usrey et al., 1999).

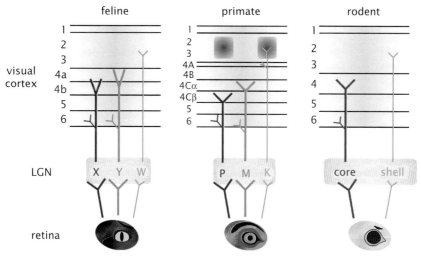

FIGURE 11.1 Schematic organization of the retino-geniculo-cortical pathway in the feline, primate, and rodent. In all mammals, including these examples, the majority of geniculo-cortical axons terminate in cortical layer 4, a minority of axons terminate in cortical layers 1–3. In the feline, there are three major parallel pathways to cortex—the X cell pathway terminates primarily in layer 4b, the Y cell pathway terminates primarily in layer 4a, and the W cell pathway terminates in layers 1–3. There are also three parallel pathways in the primate—the parvocellular (P) pathway terminates in layer 4Cβ, the magnocellular (M) pathway terminates in layer 4Cα, and the koniocellular (K) pathway terminates in the layer 1–3 blobs. In rodents, the LGN is divided into core and shell regions. Relay cells in the core project primarily to cortical layer 4, whereas relay cells in the shell project primarily to the overlying superficial layers. Adapted from Rathbun and Usrey, 2009.

The projections of X, Y, and W cells to the cortex indicates that there is both stream mixing and stream segregation in the geniculocortical pathway (Figure 11.1). Whereas geniculate W cells project to the superficial layers of V1, geniculate X and Y cells project to layer 4. Within layer 4, the axons of X and Y cells are partially overlapping, providing a possible substrate for stream mixing. Although limited, data support the notion of stream mixing as well as stream segregation to individual layer 4 cells (Alonso et al., 1996, 2001).

11.2 Primates

11.2.1 Distinct Classes of Retinal Ganglion Cells Establish Parallel Pathways in the Primate

Similar to the cat, the retina of primates contains a rich diversity of retinal ganglion cell types that differ in their morphology, physiology, and connectivity with central brain structures. Using ex vivo approaches to study the living retina of macaque monkeys, researchers have examined and compared all of these properties in the same cells and have used this information, along with information from other approaches, to compile a catalog of retinal ganglion cell types (Wässle et al., 1989; Watanabe and Rodieck, 1989; Dacey, 1993, 1999, 2000; Dacey and Lee, 1994; Martin et al., 2001; Dacey et al., 2003; Wässle, 2004; Masland, 2011, 2012; Crook et al., 2009, 2014a,b). Importantly, current understanding of retinal organization at the microcircuit and ganglion cell level is approaching a nearly complete picture, with 96.5% of retinal ganglion cells in the macaque monkey containing 19 different cell classes (Table 11.1). Many of these cell classes innervate relay cells in the lateral geniculate nucleus and therefore are the source of parallel streams of information to cortex. Notably, on- and off-center midget ganglion cells comprise approximately 58% of all retinal

TABLE 11.1 Diversity of retinal ganglion cell types in the macaque monkey

	Ganglion cell morphological type	% of total ganglion cell population*	Central projections	Some physiological Properties
1	Midget inner (ON-center)	27.0	LGN parvo 3-6	L vs M cone opponent;
2	Midget outer (OFF-center)	31.0		ON and OFF- center-surround; Achromatic/chromatic, sustained
3	Parasol inner (ON-center)	8.0	LGN magno 1, 2; Superior colliculus; Pretectum	Achromatic;
4	Parasol outer (OFF-center)	7.8		L vs M cone non-opponent; ON and OFF center surrounded; transient
5	Small bistratified (ON-OFF)	6.1	LGN konio 3, 4	S ON; L+M OFF opponent; Small and coextensive receptive field
6	Recursive monostratified	1.4	LGN; Pretectal area (NOT?)	Possible correlate of non- primate ON-direction selective cells
7	(3 inner populations)	1.4		
8		1.4		
9	Narrow thorny inner (ON-center)	1.5	LGN; Superior colliculus; Pretectum	Transient; achromatic; L+M cone inputs; ON and OFF centers, S-cone input to ON cells only.
10	Narrow thorny outer (OFF-center)	1.5		
11	Smooth inner (ON-center)	1.3	LGN; Superior colliculus	Achromatic L+M cone input
12	Smooth outer (OFF-center)	1.3		
13	Sparse inner (ON-center)	0.9	LGN; Superior colliculus	L+M ON; S OFF opponent details unknown for space outer cells
14	Sparse outer (OFF-center)	0.9		
15	Recursive bistratified (ON-OFF)	1.5	LGN; Superior colliculus	Possible correlate of rabbit ON-OFF direction; selective ganglion cells
16	Large bistratified (ON-OFF)	1.5	LGN; SC	S ON vs L+M OFF; opponent large receptive field
17	Broad thorny (ON-OFF)	1.3	LGN; Superior colliculus; Pretectum	Primate correlate of mouse and rabbit "Local edge detector" ON-OFF, sustained, achromatic
18	Melanopsin inner (ON-center)	0.5	LGN	Sustained ON response
19	Melanopsin outer (OFF-center)	0.6	Superior colliculus; Pretectal area; PON, SCN	S OFF; L+M ON opponent Strong rod input Intrinsically photosensitive Intraretinal axon collaterals

% total ganglion cell population 96.5

These cell classes differ in their morphology, physiological properties, and central targets. Approximately 84% of known retinal ganglion cells have axons that innervate the lateral geniculate nucleus (many also innervate other central structures). LGN, lateral geniculate nucleus; OPN, olivary pretectal nucleus; SC, superior colliculus; SCN, suprachiasmatic nucleus. Adapted from personal communication with Dennis Dacey.

ganglion cells and send axons to the parvocellular layers of the lateral geniculate nucleus; on- and off-center parasol ganglion cells comprise approximately 16% of all retinal ganglion cells and send axons to the magnocellular geniculate layers; and several additional classes of retinal ganglion cells, together totaling about 10% of all retinal ganglion cells, send axons to the koniocellular geniculate layers. Thus, based on these estimates, approximately 84% of retinal ganglion cells in the macaque retina contribute to the retino-geniculo-cortical projection, and this projection contains at least 13 parallel pathways that convey distinct information about different features in the visual scene (described below).

11.2.2 Parallel Processing in the Primate Lateral Geniculate Nucleus

The primate lateral geniculate nucleus contains three types of layers—magnocellular, parvocellular, and koniocellular—each of which contains on- and off-center cells, with each representing distinct parallel streams for visual processing. Years of investigation have demonstrated that the cellular morphology, visual physiology, sources of retinal input, and projections to visual cortex differ among the magnocellular, parvocellular, and koniocellular neurons (reviewed in Schiller and Logothetis, 1990; Merigan and Maunsell, 1993; Casagrande, 1994; Casagrande and Kaas, 1994; Lee, 1996; Hendry and Reid, 2000; Nassi and Callaway, 2009). Briefly, magnocellular neurons, like their retinal inputs, have larger cell bodies, larger receptive fields, faster and more transient responses to visual stimuli, faster conducting axons, and greater response gain than do parvocellular neurons and their retinal inputs. With respect to color processing, magnocellular neurons have achromatic receptive fields, whereas parvocellular neurons, at least in Old World monkeys and some New World monkeys, have receptive fields that are chromatically selective. Specifically, parvocellular receptive fields are red–green color opponent due to the spatial organization of the long- and medium-wavelength cones that establish their center–surround receptive fields (Reid and Shapley, 1992).

The projections from the magnocellular and parvocellular layers to primary visual cortex are strictly segregated (Figure 11.1). Magnocellular axons terminate in cortical layer 4Cα, whereas parvocellular axons terminate in layer 4Cβ. Thus, layer 4C stellate cells that have dendrites restricted to either layer 4Cα or 4Cβ receive stream-specific input from the lateral geniculate nucleus. However, cells beneath layer 4C, in layers 5 and 6, have apical dendrites that extend through layers 4Cα and 4Cβ and consequently have the opportunity to sample information traveling in both streams.

The koniocellular layers of the lateral geniculate nucleus are located beneath each of the magnocellular and parvocellular layers. Relay cells in the koniocellular layers are more sparsely distributed than in the magnocellular and parvocellular layers and have axons that bypass cortical layers 4Cα and 4Cβ to make synapses in the superficial layers of primary visual cortex (Fitzpatrick et al., 1983; Casagrande et al., 2007). Neurons in the koniocellular layers are a diverse group, containing multiple subclasses of cells that receive input from distinct classes of retinal ganglion cells. These classes include the small and large bistratified cells, the recursive monostratified and bistratified cells, the smooth and sparse inner cells, and the narrow and broad thorny cells (reviewed in Dacey and Packer, 2003; Wässle, 2004; Masland,

2011, 2012; Crook et al., 2014b). In contrast to magnocellular and parvocellular geniculate neurons, which have monocular responses, some koniocellular neurons have binocular responses (Cheong et al., 2013), suggesting that individual koniocellular neurons combine retinal inputs either directly or through polysynaptic circuits. Similar to parvocellular neurons, some koniocellular neurons have chromatically selective (blue/yellow) receptive fields, with input from the short-wavelength–sensitive cones being antagonistic to mixed input from the medium- and long-wavelength–sensitive cones (Dacey and Lee, 1994; Hendry and Reid, 2000).

As evidence of the diversity of circuits that establish responses in the koniocellular layers, we now have a fairly complete understanding of the full circuitry of a retinal pathway that supplies a subset of target neurons in the koniocellular layers with orientation and directional information, similar to that found in the mouse lateral geniculate nucleus (Cheong et al., 2013; Percival et al., 2014).

Importantly, like felines, differences in the visual physiology of primate geniculate magnocellular, parvocellular, and koniocellular neurons are established in the retina. Moreover, most of the features that distinguish the magnocellular, parvocellular, and koniocellular streams in the primate are similar to those that distinguish the Y cell, X cell, and W cell streams, respectively, in the feline, indicating a common evolutionary history (see the later section 11.5 below on "Evolution, Homology and Parallel Pathways").

11.3 Rodents

Research on the mouse visual system has experienced tremendous growth in recent years, largely due to the advent of molecular tools for identifying cell types, dissecting neural circuits, and manipulating neuronal activity. Although mice are generally not considered to be highly visual animals when compared with more traditional animal models used in vision research, it would be a mistake to dismiss results obtained from the mouse. Indeed, recent results from studies in mice, including those conducted in the retina and lateral geniculate nucleus, have provided valuable information that would be difficult, if not currently impossible, to obtain in other model systems.

11.3.1 Distinct Classes of Retinal Ganglion Cells Establish Parallel Pathways in the Mouse

As with most non-primate mammals, there are two major cone types in the rodent retina, one maximally sensitive to short-wavelength (blue) light and the other maximally sensitive to medium-wavelength (green) light, that supply retinal ganglion cells with visual information via intrinsic circuits that link the photoreceptors with the ganglion cells. The density and distribution of the two cone classes is not uniform across the retina (Szel and Rohlich, 1992) but has an inverse dorsal-ventral gradient that is thought to optimize the detection of visual features in the sky versus the ground (Szel and Rohlich, 1992; Applebury et al., 2000; Haverkamp et al., 2005; Baden et al., 2016; reviewed in Wernet et al., 2014; Heukamp et al., 2020). As with cats and monkeys and described in recent reviews (Masland, 2012;

Seabrook et al., 2017; Heukamp et al., 2020), there is a great diversity of retinal ganglion cell types in rodents. In mice, evidence supports the existence of approximately 33 different types of retinal ganglion cells that tile the retina (Baden et al., 2016), and these together project to approximately 40 different targets in the rest of the brain. The lateral geniculate nucleus is the primary retinal target for visual signals destined for cortex, and the diversity of retinal ganglion cell types that innervate the geniculate may be as large, if not larger, than that for the primate. Unlike the primate, where the density of parvocellular and magnocellular projecting retinal ganglion cells is far greater than the density of koniocellular projecting neurons, the density of different classes of geniculate projecting retinal ganglion cells is more balanced in the mouse, thus making the mouse a particularly useful model organism for finding and studying the ganglion cell types that are difficult to find in primates and cats. Among the retinal ganglion cells that project to the mouse lateral geniculate nucleus, there are on-center and off-center, X-like and Y-like retinal ganglion cells that project to the core region of the lateral geniculate nucleus, and W-like cells that project to the overlying shell (Seabrook et al., 2017).

11.3.2 Parallel Processing in the Mouse Lateral Geniculate Nucleus

Electrophysiological and optical recordings from the mouse lateral geniculate nucleus have revealed greater diversity among cell types and response properties than has been reported for either cats or primates. This diversity includes cells that have response properties generally considered more typical of cortical neurons in monkeys and cats, such as orientation selectivity and direction selectivity, and cells that have response properties not commonly found in the retino-geniculo-cortical pathway in more traditional animal models, such as neurons that encode the absence of contrast or the axis of motion selectivity (Huberman et al., 2009; Marshel et al., 2012; Piscopo et al., 2013; Scholl et al., 2013; see also Barlow et al., 1964). Because we know relatively little about the diversity of cell types in the cat W pathway and the primate koniocellular pathway, it may be that the cell types identified in the mouse have a counterpart in other model organisms. In support of this possibility, anatomical and physiological studies have revealed a much greater diversity of retinal ganglion cells in the primates with projections to koniocellular layers compared to those with projections to the magnocellular and parvocellular layers (reviewed in Dacey and Packer, 2003; Wässle, 2004; Crook et al., 2014b; Masland, 2011, 2012). Thus, results from the mouse may provide insight into the broader range of possible processing strategies utilized across mammals.

11.4 Parallel Pathways from the Lateral Geniculate Nucleus to Cortex Are Matched with Parallel Feedback Pathways

As discussed in Chapter 8, the lateral geniculate nucleus and primary visual cortex are reciprocally connected by feedforward and feedback projections. The majority of layer 6 neurons in primary visual cortex with axons that leave the cortex project to the lateral geniculate nucleus (note: these neurons are distinct from those that project to the claustrum; LeVay and Sherk, 1981; Kim et al., 2014, 2016; Brown et al., 2017). In addition to their geniculate projections, layer 6 corticogeniculate neurons also send axon collaterals to the overlying layers of cortex that receive thalamic input[1] (Lund and Boothe, 1975; Gilbert and Wiesel, 1979; Usrey and Fitzpatrick, 1996; Briggs et al., 2016). By projecting to the geniculate and the layers of cortex that receive geniculate input, layer 6 neurons are in a strategic position to influence communication between the geniculate and cortex.

Similar to the stream-specific organization of the feedforward projection from the retina through the geniculate to visual cortex, the corticogeniculate feedback pathway also contains stream-specific projections. This is well-documented in the monkey, where the corticogeniculate pathway includes separate classes of corticogeniculate neurons with axons selective for the magnocellular, parvocellular, and possibly even the koniocellular layers of the lateral geniculate nucleus (Fitzpatrick et al., 1994; Ichida and Casagrande, 2002; Ichida et al., 2014).

The visual responses of corticogeniculate neurons support the idea of a pathway comprised of parallel streams. Neurons in the lower tier of layer 6 in monkeys, which are known to project to the magnocellular geniculate layers, have magnocellular-like response properties: they respond well to low-contrast and fast-moving stimuli and have fast-conducting axons. In contrast, neurons in the upper tier of layer 6, which are known to project to the parvocellular geniculate layers, have parvocellular-like response properties: they prefer higher contrast and slower moving stimuli and have slower conducting axons (Fitzpatrick et al., 1994; Briggs and Usrey, 2009). Based on these properties, visual stimuli well-suited for exciting the magnocellular pathway should excite corticogeniculate neurons with axons selective for the magnocellular layers of the lateral geniculate nucleus, whereas visual stimuli better suited for exciting the parvocellular pathway should excite corticogeniculate neurons with axons selective for the parvocellular layers. Although less is known about corticogeniculate projections to the koniocellular layers of the lateral geniculate nucleus, existing evidence indicates that neurons in these layers also receive stream-specific feedback input from cortex (Fitzpatrick et al., 1994; Usrey and Fitzpatrick, 1996; Briggs and Usrey, 2009). The corticogeniculate pathway, therefore, appears poised to adjust the gain of thalamocortical communication in a stream-specific fashion. Less is known about corticogeniculate projections in other species; however, evidence from the rat, cat, ferret, and tree shrew supports the notion of parallel streams in the corticogeniculate pathway (Tsumoto and Suda, 1980; Bourassa and Deschênes, 1995; Usrey and Fitzpatrick, 1996; Briggs and Usrey, 2005).

The presence of corticogeniculate neurons with physiological properties that align with the magnocellular, parvocellular, and koniocellular streams implies that there must be circuits within the visual cortex that maintain stream-specific processing, from input to output. Consistent with this idea, it is worth noting that many corticogeniculate neurons have dendritic arborizations that branch preferentially in overlying cortical layers associated with either the magnocellular, parvocellular, or koniocellular streams (Usrey and Fitzpatrick, 1996; Briggs et al., 2016; Hasse and Briggs, 2017a).

---

1. In the rodent, layer 6 corticothalamic neurons also provide strong input to layer 5a (Kim et al., 2014).

11.5 Evolution, Homology and Parallel Pathways

Much research into the organization of thalamus and cortex has, as its underlying goal, the development of insights into the general mammalian plan for such organization. In other words, if the model animal under study is the mouse, the usual implicit goal is to direct questions aimed at general mammalian features and avoid those that are relevant only to the mouse. (This is not a straightforward strategy to implement.) Ideally, these organizational properties are studied and compared among many mammalian model species, and commonalities that emerge can be regarded as features common to all mammals. In this way, such research can also gain insight into these organizational features in humans even if actual data from humans are sparse or nonexistent.

11.5.1 Homology Versus Analogy

Attempts to determine common organizational features between species usually involve one of two approaches: *homology* or *analogy*. Homology refers to features seen as similar in related species due to their inheritance from a relatively recent, common ancestral species. An example of homologous features are our arms and hands and the wings of birds. They are closely related in evolutionary terms but clearly have evolved very different functions, such as grasping versus flying. Analogy refers to features that evolved to perform a similar function but are not clearly related in evolutionary terms. An example of analogous features are the fins of fishes and flippers of whales. These have not emerged from a recently common ancestor but have evolved for a similar purpose—namely, swimming—from disparate ancestral structures. Analogous features such as these are often referred to as *products of convergent evolution*.

The distinction between homology and analogy is quite important because establishing homologous features between species is the "gold standard" for providing insights into any mammalian master plan; thus, establishing homologous relationships for thalamocortical properties across mammalian species is key for understanding the underlying mammalian design. And thus, appreciation of the homologous relationship between our arms and hands and wings of a bird is quite informative in regards to the relationship of these structures between species. In contrast, knowing that fins in fish and flippers of whales are used for swimming offers little insight into how these structures might be related in the different species.

Indeed, the whole point of evolution is, via mutations, to alter the function of a structure in an ancestral organism to allow descendant organisms to better adapt to their complex and changing environments. Thus, a common ancestor to birds and mammals had structural features that evolved into wings in birds and forepaws in mammals. It follows that homologous features in different species may have quite dissimilar functions, and similar functionality (like fins in fishes and flippers of whales) may not be associated with homology at all.

Generally, in biology, the fossil record is quite useful in establishing homologies. However, a rather obvious point is that brains do not fossilize, and so this approach is unavailable to neuroscientists attempting to establish homologies. This has resulted in much controversy and too little certainty in homologies that have been suggested. Thus, much of what follows should be regarded with healthy skepticism.

11.5.2 Homology of Thalamic Nuclei

Most thalamic nuclei were named in multiple mammalian species by late 19th- and early 20th-century anatomists who were quite conscious of using principles of homology in this process. They based their ideas of homology largely on neuroanatomical criteria, such as location within thalamus and connectivity. In this way, the lateral geniculate nucleus was identified as a homologous structure in many species based on its location within thalamus, its innervation by retina, and its innervation of posterior cortical areas identified as visual. We accept the homology and common names for many thalamic nuclei, because we believe that these early neuroanatomists got their terminology right. Examples are the main sensory thalamic relays (e.g., the lateral geniculate, medial geniculate, and ventral posterior nuclei), the main motor thalamic relays (e.g., the ventral anterior and posterior complex), and the medial dorsal nucleus.

However, there are some examples in which the naming accuracy is open to question and needs to be reconsidered. One example is the pulvinar, which we generally recognize as present in most mammals but not in rodents. Because the pulvinar is the largest visual thalamic relay, its "absence" in rodents has been used as an argument that mice are a poor model to study vision. However, rodents, including mice, have a thalamic relay, the lateral posterior nucleus, named as such by early neuroanatomists, and more recent evidence of connectivity makes it pretty clear that the lateral posterior nucleus is indeed homologous to the pulvinar seen in other mammals (Zhou et al., 2017).

Another apparent naming problem involves the posterior medial nucleus. This is a well-defined structure in rodents and carnivores found near the ventral posterior nucleus, but its identity in primates is unclear. A thalamic structure in primates called the posterior nuclear complex has been described, and a medial region of this complex can be seen, but this region does not seem to have the same relationship to the ventral posterior nucleus as is evident in other mammalian species. Some have suggested that a better candidate in primates for homology to the posterior medial nucleus seen in other mammals is the anterior region of the primate pulvinar (Harting et al., 1972; Jones et al., 1979; Pons and Kaas, 1985).

Note that the two preceding examples for possible misidentification regarding homology, the pulvinar and posterior medial nucleus, are what we now regard as chiefly higher order relays. Early neuroanatomists got naming right for the main sensory relays, because they understood what inputs were being relayed (e.g., the retina for the lateral geniculate nucleus). However, the source of the input to be relayed for higher order relays—namely, layer 5 of cortex—was not appreciated until relatively recently, and so it is understandable that mistaken homologous relationships were created.

There are probably other examples of improperly named thalamic nuclei (and implied homologies), particularly among higher order nuclei. What is needed is a reconsideration of terminology for thalamic nuclei based on our present understanding of relevant data. A similar problem was dealt with regarding terminological issues of avian brain by convening a group of experts to form a new consensus on the topic (Reiner et al., 2004; Ebert, 2005), and perhaps a similar approach is needed for naming of thalamic nuclei.

11.5.3 Criteria for Establishing Homology Among Neuronal and Nuclear Examples

In comparing neuronal or thalamic nuclear examples between species, a key question is: What parameters should be used for establishing homology? For instance, consider a visual neuron in two species. A wide range of parameters to be used could include receptive field properties, internal membrane properties, somadendritic morphological features, connections with other regions, molecular markers, etc. How do these rank in terms of usefulness to identify homology?

The answer to that is far from clear, but a consideration of the process of evolution provides some limited guidance in this matter. The whole purpose of evolution is to employ random mutations to change the function of a structure in a "parent" organism as new closely related "offspring" organisms evolve. The result is that homologous structures in closely related species might have quite different functions: think of our hands and a bird's wings. This would suggest that looking for functional criteria to establish homology might often be misleading.

If functional criteria are relatively likely to be altered through evolution and thus be less reliable markers of homology, which criteria are better in this regard? Again, there is no clear answer to this question. However, if we go back to the example of the homology between arms and hands and wings, we see very little similarity in the different functions evolved but rather more similarity in the underlying structure. Perhaps the basic structure of neurons and their connections is a more useful parameter to evaluate for potential homology (Krahe et al., 2011) and even more so would be molecular markers that are plausibly less susceptible to evolutionary pressures (Felch and Van Hooser, 2012; Piscopo et al., 2013).

In this context, it is useful to consider a parameter quite often used for comparison of visual neurons among species: the receptive field. Attempts to establish homologies between the parallel visual processing streams in primates (the koniocellular, parvocellular, and magnocellular streams) and carnivores (the W, X, and Y streams) run into difficulties if reliance is placed on the receptive field. For example, the presence of wavelength sensitivity has been used to argue that cat W cells were homologous with monkey P cells and that monkey M cells included both X and Y homologs (Shapley and Perry, 1986); K cells were not considered here. Likewise, regarding nonlinear subunits in the receptive field, their absence (an X cell feature) or presence (a Y cell feature) has been used to find X and Y homologs in the magnocellular layers of the monkey (Kaplan and Shapley, 1982) and in the retina of the rat (Heine and Passaglia, 2011).

As an alternative, it seems entirely plausible that evolution could result in a parvocellular homolog that has wavelength sensitivity in one species but not another by a small alteration of synapses in the retina whereby cone inputs are kept separate in one species to preserve wavelength sensitivity but pooled in another, obliterating such sensitivity. Indeed parvocellular cells in the macaque monkey show wavelength sensitivity, but those in the owl monkey do not because the owl monkey lacks color vision (Jacobs et al., 1993; Tan and Li, 1999; Levenson et al., 2007; Mundy et al., 2016; Carvalho et al., 2017), which illustrates how homologous neurons in monkey species can have such different functional (e.g., receptive field) properties. Likewise, the nonlinear subunits of cat Y cells could be eliminated or suppressed by evolution in producing homologous cells that lack these. This would explain why monkey M cells that seem homologous to cat Y cells generally lack nonlinear subunits (Derrington and Lennie, 1984; Levitt et al., 2001).

Cats and monkeys share a common ancestor that existed approximately 80 million years ago, and, since that time, their evolutionary paths have led to many species-specific differences. With this in mind, it is rather remarkable how similar the distinctions between cat geniculate X and Y cells are with those of monkey parvocellular and magnocellular neurons (reviewed in Stone, 1983; Sherman, 1985; Cleland, 1986; Schiller and Logothetis, 1990; Shapley, 1992; Merigan and Maunsell, 1993). For instance, cat X cells and monkey parvocellular neurons have smaller cell bodies and slower conducting axons than do their parallel stream counterparts, the Y cells and magnocellular neurons (Schiller and Malpeli, 1978; Friedlander et al., 1981; Headon et al., 1985; Maunsell et al., 1999). In addition, the axons of cat X cells and monkey parvocellular neurons terminate in deeper regions of layer 4 than do those of Y cells and magnocellular axons (Figure 11.1; Hendrickson et al., 1978; Blasdel and Lund, 1983; Diamond et al., 1985; Freund et al., 1985; Humphrey et al., 1985a,b). Physiologically, cat X cells and monkey parvocellular neurons have smaller receptive fields and slower visual responses than do their parallel stream counterparts (So and Shapley, 1979; Troy and Lennie, 1987; Saul and Humphrey, 1990; Croner and Kaplan, 1995; Maunsell et al., 1999; Usrey et al., 1999; Usrey and Reid, 2000; Levitt et al., 2001). Cat X cells and monkey parvocellular neurons also have more sustained visual responses and prefer stimuli with lower temporal frequencies than do Y cells and magnocellular neurons (Usrey et al., 1999; Usrey and Reid, 2000; Levitt et al., 2001; Alitto et al., 2011). Last, molecular markers selective for X cells versus Y cells are also selective for parvocellular versus magnocellular neurons (Hendry et al., 1984; Hockfield and Sur, 1990; Iwai et al., 2013). Taken together, the anatomy, physiology, and molecular markers underscore a strong case that can be made for homology between cat geniculate X cells and monkey parvocellular neurons and similarly for cat Y cells and monkey magnocellular neurons.

It is worth re-emphasizing that separate from the X/Y and magnocellular/parvocellular streams to cortex which both terminate in cortical layer 4, there is an evolutionarily older pathway that terminates in the superficial layers of visual cortex (Figure 11.1). Across a broad range of species, including cats, ferrets, minks, monkeys, galagos, mice, rats, and tree shrews, the nature of the differences between the cell types projecting to cortical layer 4 versus the superficial layers is striking, suggesting that these two routes to cortex have been distinct for much of mammalian evolution. In addition to the differences in their projections to cortex, geniculate relay cells that project to the superficial layers of cortex are typically small and pale-staining, and they receive fine caliber retinal fibers (LeVay and Gilbert, 1976; Fitzpatrick et al., 1983; Weber et al., 1983; Diamond et al., 1985; Usrey et al., 1992). In addition to receiving direct retinal input, superficial layer projecting relay cells also receive visual input from the superficial layers of the superior colliculus (Graham, 1977; Fitzpatrick et al., 1980; Diamond et al., 1991), a source of input not present in relay cells with projections to layer 4. Finally, included in the great diversity of relay cells with projections to the superficial layers of cortex are cells with the largest receptive fields and slowest visual responses found in the lateral geniculate nucleus (Cleland et al., 1976; Wilson and Sherman, 1976; Sur and Sherman, 1982; Stanford et al., 1983). In cats, monkeys, and mice, relay cells that belong to the group with projections to the superficial layers of cortex are the W cells, koniocellular neurons, and shell neurons, respectively. Given their similarities with each other and their marked differences with relay cells that target layer 4, an argument can be made that these cells are homologous.

If homology exists between relay cell types, as described above, then it may seem surprising that the lateral geniculate nucleus can appear so different in terms of its shape, size, and lamination patterns (Figure 11.2). For instance, in some species (e.g., tree shrews, minks, and ferrets) the lateral geniculate nucleus has separate layers for cells with on-center versus off-center receptive fields, while in other species (i.e., cats, which are closely related to minks and ferrets) the lateral geniculate nucleus combines these cell types within the same layers. Even among primates, some species have a six-layered lateral geniculate nucleus (e.g., macaque monkey), while others have a four-layered lateral geniculate nucleus (e.g., owl monkey). Moreover, it is difficult to discern any lamination in mice. Given this high degree of diversity in geniculate layering, it is noteworthy that the visual cortex appears relatively constant in terms of its thickness and layering. This invariance in cortical morphology compared

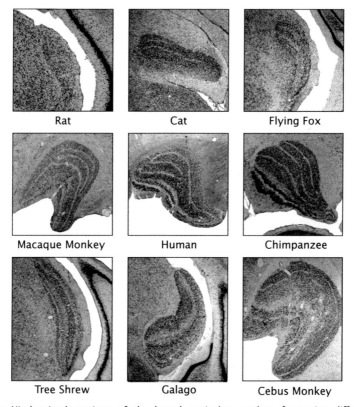

FIGURE 11.2 Nissl-stained sections of the lateral geniculate nucleus from nine different species illustrating the diversity of shapes and lamination patterns that accommodate the parallel retino-geniculo-cortical pathways. *Top row*: rat, cat, flying fox; middle row: macaque monkey, human chimpanzee; *bottom row*: tree shrew, galago (bush baby), cebus monkey. Adapted from Usrey and Alitto, 2015.

to that of the geniculate likely reflects precise developmental constraints needed to establish the much greater number of intrinsic and extrinsic circuits present in the cortex compared to the lateral geniculate nucleus.

11.6 Summary and Concluding Remarks

Across mammals, the retino-geniculo-cortical pathway is comprised of parallel streams of information flow. These streams include distinct circuits for neurons with on-center and off-center receptive fields. Further subdivisions are based on anatomical, physiological, and neurochemical properties. Importantly, the retinal ganglion cells that supply each of the parallel streams independently tile the retina and provide a complete representation of visual space. As a consequence, there are multiple maps of visual space in the optic nerve, lateral geniculate nucleus, and optic radiations. In felines, the X, Y, and W pathways share many features in common with the primate parvocellular, magnocellular, and koniocellular pathways, respectively. Although species-specific differences exist, the similarities between these cell classes greatly outnumber the differences, suggesting they represent homologous classes. The W cell/koniocellular pathway may be the most diverse, with several subclasses of cell types; however, we know relatively less about the neurons that make up these cell types. Results from studies on rodents will likely provide an avenue for understanding these cell types, as these cells comprise a larger fraction of the neurons within the retino-geniculo-cortical pathway in rodents, and new molecular techniques are providing strategies for cell identification and circuit tracing that were previously much more difficult. Less is known about parallel processing of signals en route to cortex in the other sensory modalities; however, evidence reveals afferents with features (e.g., sustained vs. transient responses, large versus small receptive fields) that distinguish cell types in the visual pathway.

11.7 Some Outstanding Questions

1. Do all thalamocortical pathways contain distinct classes of neurons similar to W, X, and Y cells of the cat's lateral geniculate nucleus?
2. If speed is generally considered a good thing, why are some pathways to cortex slower than others?
3. In cat and monkey, how many separate classes of W or koniocellular parallel streams exist, and what are their homological relationships? And what is the relationship of these to pathways in the mouse?
4. Does parallel processing of the sort seen through the lateral geniculate nucleus also occur in higher order thalamic nuclei involved in transthalamic processing?
5. In the visual system of cats and monkeys, parallel information streams are kept largely separate and independent up to at least the thalamocortical level, but to what extent are these streams maintained as separate and independent through further cortical processing? And what is the situation for parallel processing through cortex outside of the visual system?

12

Thalamocortical Substrates of Attention

Selective attention is a cognitive process that allows an organism to direct processing resources to behaviorally relevant stimuli. Selective attention serves to improve stimulus detection and discrimination of attended stimuli relative to unattended stimuli, as well as decrease reaction times for appropriate behavior. Because attention is a limited resource, the allocation of attention must be flexible and under control of the central nervous system. Neural mechanisms for attention and its allocation have been traditionally viewed as residing solely within the cortex; however, growing evidence indicates subcortical structures, including the thalamus, also play a significant role (reviewed in Halassa and Kastner, 2017; Usrey and Kastner, 2020; Sherman and Usrey, 2021). This chapter is focused on thalamocortical network interactions and their role in selective attention. Although selective attention occurs with all sensory modalities, the neural mechanisms of attention have been studied most extensively in the primate visual system, which we emphasize below. Moreover, as most studies of selective attention involving thalamus have focused on spatial attention (directing attention to specific locations), rather than feature attention (directing attention to nonspatial attributes, such as red vs. green or leftward vs. rightward motion), this chapter is focused on spatial attention.

Mechanisms underlying attention have long been a focus for considerable investigation in the field of neuroscience. A detailed discussion of attention is beyond the scope of this account, and many excellent reviews on the subject are available (Desimone and Duncan, 1995; Reynolds and Chelazzi, 2004; Maunsell and Treue, 2006; Posner, 2012; Petersen and Posner, 2012; Nobre et al., 2014). Generally, underlying mechanisms involve identifying bottom-up and/or top-down circuits that enable a brain region, typically cortical, to enhance processing of the attended object (Desimone and Duncan, 1995; Awh et al., 2012). Underlying neuronal mechanisms so far identified, among others, include enhanced responses to attended stimuli (Maunsell and Treue, 2006; Lee and Maunsell, 2010; Mineault et al., 2016; Suzuki et al., 2019); less noise correlations in firing among neurons in the attending circuit (Cohen and Maunsell, 2009); synchronization among neurons within the same cortical region (Womelsdorf et al., 2006); coherent, rhythmic neuronal firing across cortical areas (Fries, 2005; Chalk et al.,

2010; Miller and Buschman, 2013; Suzuki et al., 2019); and enhancement of thalamocortical synaptic efficacy (Briggs et al., 2013).

12.1 Effects of Spatial Attention on Visual Responses in the Lateral Geniculate Nucleus and Geniculocortical Communication

12.1.1 Studying Spatial Attention in the Laboratory

We use spatial attention throughout our daily lives for all sorts of tasks. With covert spatial attention involving vision, our eyes are directed at one location while our attention is directed elsewhere. An example of covert spatial attention is the "no-look" pass made by skilled NBA point guards. Often used during a fast break when players are running full speed from one end of the court to the other, the point guard with the ball will fool the opposing team by looking at a player in one direction while making a precise pass to a different player in a different location. For the no-look pass to work, the point guard's attention is directed to a target other than where the eyes are directed. Magic Johnson was a master of the no-look pass, which was generally followed by the phrase "it's show time" from elated fans and sports casters. Another example is seen with monkeys. Monkeys spend much of their time focusing their attention on higher-ranking individuals without making direct eye contact, which could be viewed as a threat. In the laboratory, we take advantage of the ability to fixate to (1) present stimuli at precise locations relative to the center of gaze, including locations overlapping with the receptive fields of recorded neurons, and (2) hold the stimuli at these locations for a sufficient time to measure the effects of attention on neural activity.

Figure 12.1A shows the sequence of events for a typical study of spatial attention used for humans and monkeys. With this task, the eyes fixate on a point in the center of a computer screen while multiple stimuli are shown at various preselected spatial locations. Based on a cue, the participant is informed about which stimulus location they should direct their attention toward in anticipation of a change in the stimulus, which they report to receive a reward. Importantly, in a minority of trials, the cue is misleading and directs attention away from the stimulus that will change. For both humans and monkeys, reaction times and response accuracy are faster and better when the cue is valid compared to when it is not valid (Figure 12.1B). For electrophysiological studies, one of the stimuli is typically positioned over the receptive field of a recorded neuron so that neural activity can be compared when the attention is directed toward and away from the neuron's receptive field.

12.1.2 Attention Augments Visual Responses in the Lateral Geniculate Nucleus

Within the cortex, spatial attention typically increases neuronal responses to visual stimuli at attended locations (reviewed in Reynolds and Chelazzi, 2004; Maunsell, 2015) and increases coherence between single-unit activity and the local field potential in specific frequency bands (reviewed in Fries, 2015; Buschman and Kastner, 2015). Although the effects

(A)   Spatial Attention Task

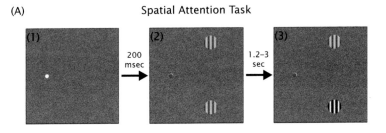

(B)   Behavioral Performance

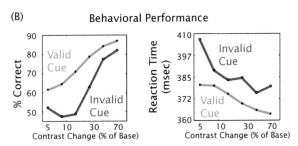

FIGURE 12.1 Example of a spatial attention task and the behavioral effects of attention. A. Attention task, sequence of events: (1) A fixation point appears and the animal obtains fixation; (2) two drifting gratings appear; the fixation point changes color 500 msec following their appearance to indicate which grating ("target") the animal should attend to while maintaining fixation; (3) following a random delay (1–3 sec), the contrast of the attended grating changes (95% trials, valid cue). The animal is rewarded for reporting the contrast change by making a saccade to the stimulus that changed. In 5% of the trials ("catch" trials), the contrast of the unattended grating changes. B. Animals are more likely to detect the stimulus change (left panel) and are faster at reporting the change (right panel) when attention is directed towards the stimulus that changed (valid cue trial) compared to the stimulus that did not change (invalid cue trial).

of spatial attention are usually strongest in extrastriate cortical areas (e.g., V4, MT, VIP), attention can also influence neuronal activity in primary visual cortex and in the lateral geniculate nucleus (O'Connor et al., 2002; McAdams and Reid, 2005; McAlonan et al., 2006; Chen et al., 2008; McAlonan et al., 2008; Schneider and Kastner, 2009; Lovejoy and Krauzlis, 2010; Briggs et al., 2013; Krauzlis et al., 2018; Hembrook-Short et al., 2019). For instance, spatial attention increases the single-unit activity of lateral geniculate neurons in macaque monkeys (Figure 12.2A) and the blood oxygenation level–dependent (BOLD) response measured during functional magnetic resonance imaging (fMRI) experiments in the human lateral geniculate nucleus (O'Connor et al., 2002; McAlonan et al., 2008; Schneider and Kastner, 2009). The mechanisms contributing to the effects of attention on neurons in the lateral geniculate nucleus likely include the release of inhibition from the thalamic reticular nucleus as spatial attention decreases the activity of thalamic reticular neurons (McAlonan et al., 2000, 2006; see also Wimmer et al., 2015). Because the influence of attention on neurons in the thalamic reticular nucleus is more transient than that for neurons in the lateral geniculate nucleus, additional pathways and mechanisms seem necessary for the full manifestation of attentional effects in the lateral geniculate nucleus. Feedback from the cortex is one likely candidate for the extended effects of attention on lateral geniculate

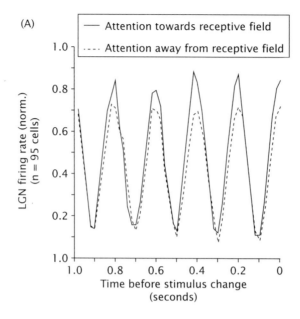

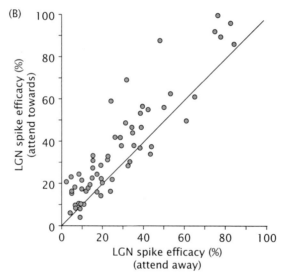

FIGURE 12.2 Influence of attention on neuronal activity in the lateral geniculate nucleus and on geniculocortical communication in the macaque monkey. A. The firing rate of LGN neurons is greater when attention is directed towards their receptive fields than when attention is directed away. The plot shows the average firing rate of 95 LGN neurons performing the contrast-change detection task, similar to that in Figure 12.1. The sinusoidal pattern of firing rate changes reflects the changes in luminance with the drifting sinusoid stimulus. Adapted from Usrey and Alitto, 2015. B. Synaptic communication between LGN neurons and target neurons in layer 4C of V1 is enhanced with spatial attention. Here, animals perform a similar attention task to that described for Figure 12.1; however, an electrical stimulating electrode placed into the LGN evokes spikes at specific times while animals attend towards or away from the RF of a synaptically connected cortical layer 4C neuron. The efficacy of electrically evoked geniculate spikes to evoke a cortical response (i.e., % of successful shocks) is shown when animals attend towards and away from the receptive field of the recorded cortical neuron. Adapted from Briggs et al., 2013.

neurons. If so, then the parallel streams of corticogeniculate projections (Chapter 11) that selectively innervate the magnocellular, parvocellular, and koniocellular layers of the lateral geniculate nucleus (Fitzpatrick et al., 1994; Briggs and Usrey, 2009; Ichida et al., 2014; Briggs et al., 2016) may provide a substrate for feedback to exert stream-specific attentional effects on visual signals traveling from the lateral geniculate nucleus to cortex.

12.1.3 Attention Augments Geniculocortical Communication

In addition to augmenting visual responses in the lateral geniculate nucleus, spatial attention also modulates the strength of geniculocortical communication. This effect was revealed by pairing electrical stimulation of neurons in the lateral geniculate nucleus with recordings from synaptically coupled target neurons in macaque V1. With this preparation, spatial attention increases the percentage of electrically evoked spikes that successfully drive postsynaptic responses in V1 (Figure 12.2B; Briggs et al., 2013). Thus, attention not only increases the firing rate of neurons in the lateral geniculate nucleus, but it also increases the efficacy, or likelihood, that geniculate spikes are successful in evoking postsynaptic cortical responses.

Rhythmic (also called oscillatory) activity patterns are common throughout the brain and have been proposed to play a role in facilitating neuronal communication between brain regions that oscillate in phase with each other (Gray and Singer, 1989; Fries, 2005). With this mechanism in mind, it is interesting to note that oscillatory phase synchronization between the local field potentials measured in the lateral geniculate nucleus and V1 has been reported for the α (8–14 Hz) and β (15–30 Hz) frequency bands (Bastos et al., 2014). An analysis of directed connectivity further reveals that β-band interactions are mediated by geniculocortical feedforward processing, whereas α-band interactions are mediated by corticogeniculate feedback processing (Bastos et al., 2014). Given the presence of oscillatory activity in the lateral geniculate nucleus and V1 along with the phase synchronization measured between the two structures, an important question to answer is whether or not attention can modulate the strength of the oscillatory interactions between the two structures, as has been shown to occur with the pulvinar and cortex (described below).

12.2 Effects of Spatial Attention on Visual Responses in the Pulvinar and Pulvino-Cortical Network Interactions

Compared to other thalamic nuclei, the pulvinar has undergone the greatest expansion in size with primate evolution. As a result, the pulvinar is the largest nucleus in the primate thalamus, many times the size of the lateral geniculate nucleus. The pulvinar is also considered a higher order thalamic nucleus because much or most of its driving input comes from layer 5 of cortex, rather than from subcortical sources. Pulvinar neurons, in turn, project back to cortex, both in a feedforward and feedback fashion, as described in Chapters 6 and 9.

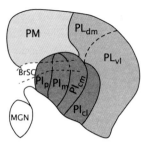

FIGURE 12.3 The organization of the anthropoid pulvinar (coronal view). The three shades of grey represent the main divisions of the pulvinar. Light grey indicates the location of the medial pulvinar (PM); medium gray indicates the location of the lateral pulvinar (PL); and dark grey indicates the location of the inferior pulvinar (PI). Within the lateral pulvinar, there are at least two divisions, the lateral pulvinar dorsal medial (PLdm) and the lateral pulvinar ventral lateral (PLvl) divisions. The inferior pulvinar contains four divisions: the posterior nucleus of the inferior pulvinar (PIp), the medial nucleus of the inferior pulvinar (PIm), the central medial nucleus of the inferior pulvinar (PIcm), and the central lateral nucleus of the inferior pulvinar (PIcl). Dorsal is up, lateral is to the right. Adapted from Kaas and Baldwin, 2019.

Broadly speaking, the anthropoid pulvinar can be divided into three divisions: the medial, lateral, and inferior pulvinar (Figure 12.3; Adams et al., 2000; Kaas and Baldwin, 2019). The inferior and medial pulvinar are located ventrally and dorsally, respectively, whereas the lateral pulvinar has both a ventral and a dorsal part. These divisions of the pulvinar have distinct physiology and connectivity patterns with the cortex. Among them, the inferior and ventral subdivisions of the lateral pulvinar have the highest number of visually responsive neurons and likely more than one retinotopic map (Kaas and Lyon, 2007; Arcaro et al., 2015), and the dorsal subdivisions of the lateral pulvinar and the medial pulvinar contain the lowest number of visually responsive neurons. Altogether, the pulvinar is likely unsurpassed among thalamic nuclei in terms of the number of cortical areas it is associated with via a diverse network of cortico-pulvino-cortical circuits. These networks likely play a major role in the communication of signals between cortical areas, as described in Chapter 6. Here, we describe additional roles for these circuits during attention.

12.2.1 Pulvinar Lesions Disrupt Attention

Pulvinar lesions in humans and monkeys provide compelling evidence that the pulvinar plays a prominent role in visual attention. Several deficits of attention are associated with thalamic lesions that include the pulvinar, including *visuo-spatial hemi-neglect*, a syndrome that includes difficulty in directing attention to objects in the contralateral visual field. Cortical lesions that include the posterior parietal cortex are also associated with a similar deficit in directing attention to the contralateral visual field. Given these similar deficits, it is noteworthy that posterior parietal cortex is interconnected with the dorsal pulvinar and that reversible inactivation of the dorsal pulvinar in monkeys results in deficits with directing attention to contralateral visual space (Wilke et al., 2010).

Another deficit of attention associated with pulvinar lesions in patients is the ability to filter distracting information. For these patients, performance in discriminating target stimuli is impaired when the target stimulus is presented along with non-target stimuli (i.e., distractors), presumably because the distractors compete with the target for attentional resources (e.g., Snow et al., 2009). Humans with cortical lesions in posterior parietal cortex and/

or V4 display a similar deficit in filtering distractor stimuli (Gallant et al., 2000; Friedman-Hill et al., 2003), as do monkeys with lesions that affect area V4 (De Weerd et al., 1999). The similarity of these deficits following cortical and pulvinar lesions suggests that the network of pulvino-cortical circuits plays a critical role in attention, because lesions to either node in the network have similar consequences on attention.

12.2.2 Attention Modulates Neuronal Activity in the Pulvinar

Electrophysiological and fMRI studies report increased activity in the pulvinar with tasks involving attention. Among the fMRI studies, results from humans show increased activity in the pulvinar with attention directed to specific locations (e.g., Arcaro et al., 2018), filtering of distracter information (e.g., Fischer and Whitney, 2012), and shifts of attention across the visual field (e.g., Yantis et al., 2002). In monkeys, electrophysiological studies show that spatial attention modulates visual responses in the dorsal, lateral, and inferior divisions of the pulvinar (Petersen et al., 1985; Saalmann et al., 2012; Zhou et al., 2016). In addition to augmenting response magnitude, attention also reduces neuronal response variability in the pulvinar and, therefore, the variability of signals communicated to cortex (Petersen et al., 1985; Saalmann et al., 2012).

12.2.3 Attention Augments Pulvino-Cortical Interactions

Studies examining cortico-cortical functional interactions suggest that synchrony in oscillatory activity between cortical areas facilitates the successful communication of signals between the areas (reviewed in Buschman and Kastner, 2015; Fries, 2015). With this idea in mind, researchers have examined whether the pulvinar might play a role in synchronizing oscillatory cortical activity as well as whether attention plays a role in the process. In these technically demanding experiments, simultaneous recordings were made from two interconnected cortical areas, V4 and TEO, as well as from the division of the pulvinar connected with both regions in the pulvinar of macaque monkeys while they performed a spatial attention task (Saalmann et al., 2012). Results revealed cortical areas V4 and TEO synchronized in the α frequency range and, to a smaller extent, in the γ frequency range when animals engaged in the spatial attention task. Results further revealed that the pulvinar causally influenced the oscillatory activity in both V4 and TEO in the α frequency range (Figure 12.4) and perhaps also in the γ frequency range, supporting the idea that the pulvinar can coordinate oscillatory activity between cortical areas with attention and therefore regulate information transfer based on attentional demands.

12.3 An Evolutionary Perspective on Attention

12.3.1 Why Does Attention Reduce Cognitive Abilities to Unattended Objects?

The explanation for attentional mechanisms is that it enables our brains to focus on environmental events of particular importance to our survival. For instance, a rabbit traveling through bushland might attend with its visual system on the lookout for hovering hawks. However, attention comes at a price, because that rabbit, by emphasizing visual stimuli, may be less responsive

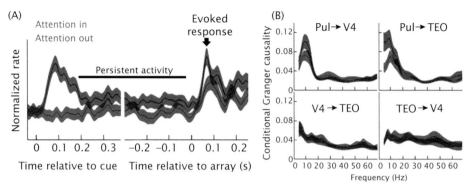

FIGURE 12.4 Attentional modulation in the pulvinar of the macaque monkey. A. When attention is directed to a pulvinar neuron's receptive field by a visual cue ("attention in") as compared to when attention is directed away from it ("attention out") there is an increase in persistent activity during a delay and moderate attentional enhancement in response to an array. B. During the delay period of an attention task, conditional Granger causality analysis indicates a role for the pulvinar in increasing coherence in an alpha frequency band between V4 and TEO. Areas V4 and TEO do not appear to interact via the direct cortico-cortical pathways during that period; instead, inter-areal interactions appear to go mainly through the indirect transthalamic pathways. Adapted from Saalmann et al., 2012.

to auditory cues that could signal a stalking fox. Even within vision there may be a price to be paid: by concentrating on upper visual fields where hawks fly, the rabbit might miss observing the fox in its lower visual field. This raises the question: Given the extensive cortical circuitry subserving its enormous computational power, why cannot all areas of cortex function in an attentive-like mode so that the rabbit can be maximally sensitive to all sensory stimuli simultaneously? We believe that an evolutionary perspective offers a plausible answer to that question (see also Sherman and Usrey, 2021).

The cerebral cortex evolved subsequently to and then alongside the evolution of lower motor circuits to which it has access. However, as cortex evolved, there was no co-evolution of a motor plant to which cortex has unique access. Moreover, cortex can only influence behavior by projections from layer 5 neurons that activate the older motor centers in the brainstem and spinal cord. These older motor centers may be seen as a bottleneck through which cortex must operate. This presents a problem. If, as suggested above, every cortical area operated at maximum capacity to turn its inputs into layer 5 motor commands, these would all compete for control through subcortical intermediaries, and chaos would likely ensue. Thus, there needs to be some selective process that ensures that only those cortical areas engaged in analyzing environmental objects that are most important, or most crucial to survival, are permitted to control subcortical motor centers. This is where "attention" comes in (Sherman and Usrey, 2021). Somehow, via top-down or bottom-up processes (Desimone and Duncan, 1995; Awh et al., 2012), the appropriate region or regions of cortex are engaged and their layer 5 corticofugal projections are allowed to dominate subcortical motor regions; other areas of cortex (and their layer 5 outputs) dealing with less critical environmental events are suppressed.

12.3.2 Attention Is Not Just Cortical

As mentioned at the outset of this chapter, most studies on attention are concerned only with cortical contributions thereof. However, an evolutionary perspective suggests a more complex view. Just as attention seems necessary to ensure that the appropriate cortical regions take control of behavior, more primitive species had to deal with the same problem but without cortex. For example, the highest level of behavioral control for nonmammalian vertebrates would be various brainstem motor areas, and, as we argue in Chapter 13, the highest level of control affecting behavior for these species is located in the midbrain. But these centers, like mammalian cortex, also had to operate through older brainstem and spinal centers, and the same problem as suggested above had to be overcome: that is, to avoid chaos, something like attentional mechanisms would be required to filter out inappropriate midbrain centers from controlling behavior (Sherman and Usrey, 2021).

One general rule of evolution of our nervous system is that these older circuits are not discarded as newer ones evolve, and these older circuits continue to function. Indeed, for example, the mammalian superior colliculus has been shown to be involved in various attentional processes (Krauzlis et al., 2013; Basso and May, 2017; Suzuki et al., 2019; Basso et al., 2021). It thus seems likely that attentional mechanisms in our brains are not limited to cortical circuitry but involve older, subcortical circuits as well, and these all must operate in a coordinated fashion.

12.4 Summary and Concluding Remarks

With attention, the cognitive resources of the brain are directed to specific stimulus attributes, such as location, color, shape, and motion. The ability of the brain to selectively focus attention has a direct impact on an organism's success and survival. Although attention has traditionally been viewed as a cortical phenomenon, increasing evidence indicates the thalamus and thalamocortical interactions contribute to attentional processes. Studies primarily with non-human primates have provided new insights into the role of the thalamus in visual attention. As with cortical neurons, neurons in the lateral geniculate nucleus and the pulvinar increase their activity with attention. This effect is stronger in the pulvinar than in the lateral geniculate nucleus, and mechanisms likely include increased excitation from the cortex along with disinhibition from the thalamic reticular nucleus. The pulvinar may also play a more direct role in mediating the effects of attention by synchronizing rhythmic activity between cortical areas, presumably to enhance communication between areas. Clearly there is much that remains to be learned about how the thalamus and thalamocortical interactions contribute to attention.

Neuronal correlates to attention are not limited to thalamus and cortex but are also evident in earlier evolved subcortical centers, such as the midbrain. We have also argued that a major purpose for attentional mechanisms seen at cortical levels is to make certain that those cortical areas involved in the most important analyses of the environment take control of older evolved subcortical centers via layer 5 projections, because these subcortical circuits represent a bottleneck between cortical messages and the motor output.

Finally, as a general rule, we gain a deeper understanding of brain mechanisms by broadening both the species and the systems being investigated. It therefore makes sense to apply a similar strategy to the study of attention.

12.5 Some Outstanding Questions

1. What are the circuits and mechanisms that underlie increased activity in the thalamus with attention?

2. Does attention utilize the parallel streams of corticothalamic feedback to direct effects to specific stimulus features (e.g., color vs. motion)?

3. Do other higher order nuclei play a role in synchronizing rhythmic activity of cortical areas with attention? Does this synchrony facilitate communication between areas?

4. Attention is able to seamlessly follow stimuli that move across cortical representations in one hemisphere to the other. Does this ability restrict possible roles for the thalamus in attention?

5. How do cortical and collicular attention mechanisms interact? Does the pulvinar play a prominent role in these interactions?

Corticothalamic Circuits Linking Sensation and Action

The evolution of cortex has provided mammals with more flexible control of behavior, thereby improving the chances of the organism to survive complex environmental challenges. Cortex, via commands sent via its layer 5 projections to brainstem and spinal motor centers, affects behavior. But necessarily linked to these motor commands are efference copies.[1] Efference copies are messages sent from motor areas of the brain back into appropriate sensory processing streams to anticipate impending self-generated behaviors. This is an absolute requirement to allow the organism to disambiguate self-generated movements from external events.

An example involving eye movements serves to illustrate the purpose of efference copies. Under most viewing conditions, we make saccadic eye movements three or more times per second (Findlay and Gilchrist, 2003). This means that, for each saccade, visual stimulation of the retina produces a signal sent to the rest of the brain that the visual world is spinning in the opposite direction of the eye movement. But we do not perceive an unstable visual world rotating several times per second. This is because a copy of the message sent to the oculomotor muscles to move the eyes is sent back into early visual processing and is used to negate the resultant visual messages sent out by the retina. We thus experience a stable world. This copy of the motor message is the efference copy.

Efference copies are a key part of brain mechanisms that allow us to distinguish apparent environmental changes caused by our own actions from actual changes in the environment. Knowing the difference between an environment that actually changes and one that remains stable when we move is required for survival of any organism that moves through the world. Indeed, coordinated behavior by any reasonably complex animal without efference copies is improbable.

---

1. The concept of *efference copy* can be compared to that known as *corollary discharge*. The concepts are quite similar, and often the terms are used interchangeably. For a comparison of some of the subtle differences between them, see Crapse and Sommer (2008a).

This point was understood by Helmholtz, who, in 1866, deduced that efference copies must exist (Helmholtz, 1866), but it was not until 1950 that efference copies were independently demonstrated experimentally in fishes (Sperry, 1950) and flies (von Holst and Mittelstaedt, 1950). This demonstration in such disparate ancient species suggests that the development of efference copies is part of our early evolutionary heritage (see below discussion) and must occur widely in the animal kingdom. It is important to note that this process requires a prediction, or a "forward model," of what will occur as a result of the impending action and that any sensory feedback that can indicate the position of the eyes or finger joints would occur *after* the movement and be too late for this purpose (Sommer and Wurtz, 2008).

Figure 13.1 shows a highly schematic version of an efference copy. An exact copy of a motor command is fed back into early stages of sensory processing to inform the sensory effects of the impending motor action. An important point is that sensory input from muscle or joint receptors resulting from the motor action would be too late to impact early sensory processing. Actual plausible circuits involved in generating efference copies are considered below.

Excellent recent reviews on the subject of efference copy are available (Crapse and Sommer, 2008a,b; Sommer and Wurtz, 2008; Wolpert and Flanagan, 2010), and so details of the subject are omitted here in order to concentrate on possible circuits, including those involving thalamus and cortex, that subserve efference copies. For purposes of this monograph, we consider the efferent projections of any neuron that strongly innervates

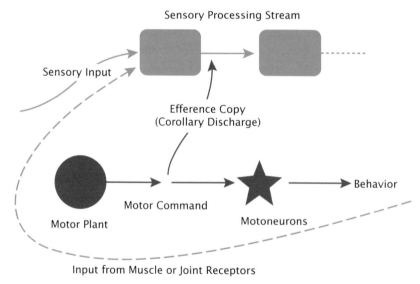

FIGURE 13.1 Schema illustrating concept of efference copy. As a motor command is sent from some source in the central nervous system (the *motor plant*) to motoneurons (*bottom blue path*), a copy of the command, which is the efference copy, is sent into early sensory processing circuitry that would be affected by the new movement. The information in the efference copy is needed for the sensory processing to account for the effects of the anticipated movement on sensation. Note that information from sensory receptors responding to the new movement (*dashed green arrow*) would arrive *after* the movement and be too late to allow early sensory processing to account for the effects of the anticipated movement. See text for details.

motor structures, such as the rubrospinal pathway, to have the opportunity to influence motor behavior. By "strongly innervate," we mean via a glutamatergic driver input rather than any modulatory one. The message that this carries is thus a potential motor message. Consequently, if these efferent projections branch to innervate additional structures, then we consider these branches to carry a copy of that motor message delivered to the motor pathway, and this copy is then what we refer to as an efference copy. Thus, efference copies need not arise solely from neurons directly innervating motoneurons. Rather, they can also arise from neurons that strongly influence motoneurons through intermediaries.

We include all cortical areas as potential sources of motor messages via their layer 5 outputs. To some readers, it may seem odd that this includes "sensory" cortical areas since many might think that sensory cortex should not have explicit motor functions. Our reasoning for this is explained below and in the chapter summary.

13.1 Driver Inputs to Thalamus Arrive via Branching Axons

Effective efference copy requires an accurate copy of a motor message to be directed into the appropriate sensory processing stream. Arguably, the best way to ensure that an exact copy of a message is transmitted from one neuron to multiple targets is via a branching axon (Cox et al., 2000; Raastad and Shepherd, 2003). That is, the same temporal pattern of action potentials will be conducted along all branches of the axon to its terminations (Figure 13.2). This does not mean that the message has the same effect on all of its targets because different synaptic properties at different targets likely exist and lead to postsynaptic variation in responses. Nonetheless a branching axon is the most efficient way to share a single message with multiple targets. This suggests a plausible role for branching axons in efference copy circuits.

13.1.1 Branching Axons and Efference Copies

Branching axons are a ubiquitous feature in the central nervous system. This may serve a general purpose of spreading a single message to different circuits, but it seems plausible that one purpose of some of the branching is to subserve the establishment of efference copies. None other than the great neuroanatomist Cajal emphasized the presence of branching axons in his Golgi-stained material and, in so doing, described patterns that in the present context could be viewed as plausible examples of efference copy circuitry.

One such example is shown in Figure 13.3A from Cajal (1911). Here, Cajal emphasized that most or all primary afferents entering the spinal cord branch, with one branch targeting the spinal gray matter, in some cases the region of motoneurons in the ventral horn, and the other branch ascending toward targets in the brain. Figure 13.3B shows a modern textbook view of the relevant circuitry here. The branch carrying the message toward motoneurons can be considered a motor command, but the branch ascending to the brain carries the exact same message. This ascending message is conventionally thought of as a sensory message, conveying information about a change in a joint angle, skin depression, etc. But because the

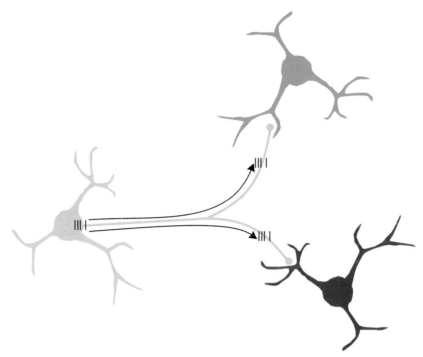

FIGURE 13.2 The message sent from a neuron (*blue*) down a branching axon would provide an exact duplicate of the temporal pattern of action potentials down all branches to arrive at different targets (*red and green neurons*). Whereas the same message is provided to each of the terminals of the blue axon, to the extent that there are different synaptic properties at the different axonal terminals, the same messages would have different postsynaptic effects.

message is an exact copy of a message targeting motor neurons and thus affecting motor behavior, it can also be seen as an efference copy.

An important point to emphasize is that the singular ascending message can be interpreted by different postsynaptic targets to different effects. As an example, consider the situation in which an army group commander sends a message of military intelligence to his soldiers on the front line. These soldiers are arranged in three groups: a center and two ends. The singular message is that the enemy is planning an encircling pincer attack by drawing resources from its center to outflank the troops on both sides. This is the only message. This one message causes the commanders of the end groups to deploy for defensive action, whereas the commander in the center decides to attack based on the information that the enemy's center will be weakened. The point is that the single message leads to quite different responses by different recipients. The point regarding Figure 13.3B is that the single message ascending to the brain via the branching axon can be interpreted by some target circuitry as sensory information and by others as an efference copy.

13.1.2 Branching Retinogeniculate Axons

It is clear that retinal axons innervating the lateral geniculate nucleus commonly branch to innervate the midbrain as well. Figure 13.4A shows an example from the cat that branches to

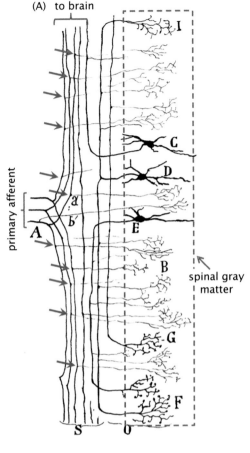

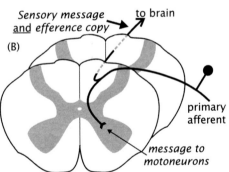

FIGURE 13.3 Cajal's view of branching primary afferent axons entering the spinal cord. A. Annotated version of Cajal's drawing (Cajal, 1911). He showed that each of these axons entering the spinal cord branches, with one branch heading to the spinal gray matter, and the other, to the brain. The *red arrows* indicate branch points. B. Modern textbook view of A. The axonal branch targeting the gray matter can be viewed as providing a message to motoneurons, which can thus be viewed as a motor message. The ascending branch carries an exact copy of that message to the brain. This one message can be interpreted by some postsynaptic targets as a sensory message (e.g., about a change in joint angle or skin indentation) and by others as a copy of a motor message, which is the definition of an efference copy. See text for details.

(A)

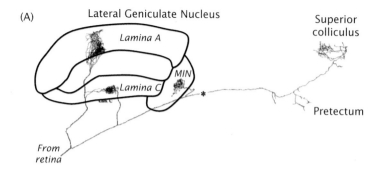

Lateral Geniculate Nucleus

Lamina A

MIN

Lamina C

Superior colliculus

Pretectum

From retina

(B)

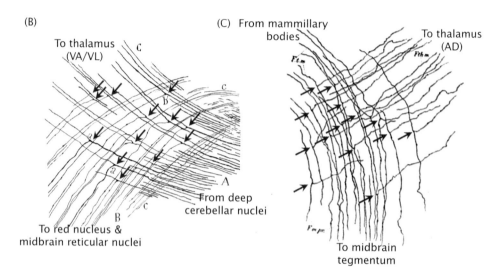

To thalamus (VA/VL)

(C) From mammillary bodies

To thalamus (AD)

From deep cerebellar nuclei

To red nucleus & midbrain reticular nuclei

To midbrain tegmentum

(D)

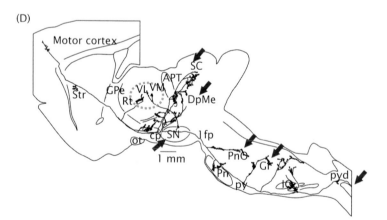

Motor cortex

SC

APT

Str

GPe

VL VM

Rt

DpMe

ot

cp SN

l fp

PnO

Gi

pyd

1 mm

Pr

py

tc

FIGURE 13.4 Examples of branching axons innervating thalamus and extrathalamic motor targets. A. Example from retinogeniculate axon of cat; redrawn from (Tamamaki et al., 1994). B. Example of cerebellar inputs to the ventral anterior and ventral lateral nuclei (VA/VL) with branches to the red nucleus and midbrain reticular nuclei (red arrows indicate branch points); redrawn from (Cajal, 1911), and thanks to Javier deFelipe for providing this image. C. Example of mammillary inputs to the anterior dorsal nucleus (AD) with branches to the midbrain tegmentum (red arrows indicate branch points); redrawn from

innervate the lateral geniculate nucleus and superior colliculus plus pretectum in the midbrain (Tamamaki et al., 1995).

The obvious question is: How representative is the example shown in Figure 13.4A? The reliable data on this point are unfortunately limited because of the difficulty in obtaining convincing evidence of branching axons. There are three common ways to identify these. First, one can do so anatomically by placing differently tagged retrograde tracers in presumed target sites (e.g., the lateral geniculate nucleus and midbrain) and look for double-labeled neurons at the afferent site (e.g., the retina). The problem here is that negative results are unconvincing because of failure either of retrograde transport or to target properly the terminal zones. Second, one can do so physiologically by recording from an afferent cell and antidromically activating it from different sites. Again, negative data are unconvincing because of plausible failure to stimulate correct target sites. Third, one can demonstrate this anatomically by Golgi impregnation or filling a single cell with dye and orthodromically tracing the axon, looking for branch points. This is the most reliable approach, although negative data would still be difficult to evaluate because both Golgi impregnation and dye filling may fail to label branches. Also, this third option is technologically the most challenging, and thus useful examples of branching axons from this approach are quite limited.

The branching of retinogeniculate axons exemplifies the issues raised above. In the cat, evidence based on retrograde labeling or antidromic activation indicates that most or all Y cells branch to innervate the lateral geniculate nucleus and midbrain, but the story for X cells is somewhat confusing, with estimates of the percentage of X cells that project to the midbrain varying widely, from about 1% (Leventhal et al., 1985) to about 10% (Wässle and Illing, 1980) to about 50% (Sawai et al., 1985). These differences in numbers should raise alarms because one might question why only a subset of a neuronal class would project to a target; a different explanation is that all X cells branch to innervate the midbrain, but many fail to be detected with retrograde labeling or antidromic activation. Indeed, studies involving intracellular labeling with dye indicate that every X and Y axon in the cat branches to innervate both the lateral geniculate nucleus and midbrain (Sur et al., 1987; Tamamaki et al., 1995). Of particular interest regarding the dye-filled axons is the observation that branching of Y axons produces thick branches targeting both the lateral geniculate nucleus and midbrain, whereas that of X axons produces thick branches to the lateral geniculate nucleus but noticeably thin ones targeting the midbrain (Sur et al., 1987; Tamamaki et al., 1995). This latter observation

---

FIGURE 13.4 Continued

(Kölliker, 1896). D. Example from layer 5 pyramidal tract cell of rat motor cortex; redrawn from Kita and Kita (2012) tracing of reconstruction generously supplied by H. Kita. Branches innervating thalamus are indicated by the dashed blue circle, and brainstem motor regions are indicated by red arrows.

Abbreviations: APT, anterior pretectal nucleus; cp, cerebral peduncle; DpMe, deep mesencephalic nuclei; Gi, gigantocellular reticular nucleus; GPe, Globus pallidus external segment; IO, inferior olive; lfp, longitudinal fasciculus of the pons; MIN, medial interlaminar nucleus (which is part of the lateral geniculate nucleus in carnivores); ot, optic tract; Pn, pontine nucleus; PnO, pontine reticular nucleus, oral part; py, medullary pyramid; pyd, pyramidal decussation; Rt, thalamic reticular nucleus; SC, superior colliculus; SN, substantia nigra; Str, striatum; VL, ventrolateral thalamic nucleus; VM, ventromedial thalamic nucleus.

might explain why, in X axons, there is a failure of antidromic transmission and retrograde transport of label through this branch point.

This, in turn raises questions about the retinofugal projection pattern in the monkey for parvocellular and magnocellular axons. Antidromic activation and retrograde labeling in the monkey indicate that magnocellular ganglion cells regularly branch to innervate both the lateral geniculate nucleus and the midbrain but that parvocellular cells innervate only the lateral geniculate nucleus (Bunt et al., 1975; Schiller and Malpeli, 1977; De Monasterio, 1978; Leventhal et al., 1981; Perry and Cowey, 1984; Rodieck and Watanabe, 1993; Crook et al., 2008). There are two possible explanations here. One is that parvocellular axons innervate only the lateral geniculate nucleus. This would make this projection relatively unique because studies in many other species (in addition to the cat noted above) indicate that the vast majority of or all retinal ganglion cells innervate the midbrain (Chalupa and Thompson, 1980; Vaney et al., 1981; Linden and Perry, 1983; Dreher et al., 1985), implying that those that innervate the lateral geniculate nucleus must be branches. This may be the case because, unlike other retinal ganglion cell axons that innervate the midbrain, parvocellular retinofugal projections may not carry a motor message. The second is that failure to find evidence of a parvocellular projection to the midbrain in monkeys reflects the sort of negative artifact discussed above.

In any case, it seems clear that, with perhaps some exceptions, the norm for retinogeniculate axons is that they branch to innervate the midbrain as well.

13.1.3 Other Branching First Order Afferents

Relevant evidence for branching axons innervating first order thalamic targets other than the lateral geniculate nucleus is quite rare and scattered but, when present, suggests a pattern of afferent branching with extrathalamic targets. These examples have been previously discussed (Sherman and Guillery, 2006) and are briefly noted here. Axons in the spinothalamic tract and medial lemniscus innervating the somatosensory thalamus branch to innervate many targets in the brainstem (Berkley, 1975; Berkley et al., 1980; Feldman and Kruger, 1980; Bull and Berkley, 1984; Lu and Willis, Jr., 1999). Figure 13.4B shows that cerebellar axons that innervate the ventral anterior/lateral thalamic complex branch (red arrows show the branch points) to innervate the red nucleus and midbrain reticular nuclei (Cajal, 1911). Figure 13.4C shows that mammillothalamic axons branch (again, red arrows show the branch points) to innervate the anterior dorsal nucleus of the thalamus as well as the midbrain tegmentum (Koelliker, 1896).

Much more evidence is needed concerning the branching pattern or lack thereof of driving afferents to first order thalamic nuclei but, where present, relevant data indicate that branching is common and perhaps even omnipresent for these afferents.

13.1.4 Branching of Layer 5 Corticothalamic Afferents

Although the data are scattered and limited, where available, they point to a pattern in which every layer 5 corticothalamic axon that innervates the thalamus branches to innervate extrathalamic sites as well (Casanova, 1993; Deschênes et al., 1994; Bourassa and Deschênes,

1995; Bourassa et al., 1995; Rockland, 1998; Guillery et al., 2001; Kita and Kita, 2012; Sherman and Guillery, 2013; Sherman, 2016). Figure 13.4D shows an example of this pattern.

13.2 Do Branching Driver Inputs to Thalamus Serve as Efference Copies?

A consequence of branching of driver inputs to thalamus is that the message sent through thalamus to cortex is an exact copy of a message sent to one or more extrathalamic sites. A remarkable feature of these extrathalamic sites is that, in every case so far seen, some of them seem to be motor centers. Examples can be seen in Figure 13.4. Figure 13.4A shows that the extrageniculate branch innervates the superior colliculus and pretectum, midbrain sites involved in the control of eye movements, pupillary constriction, and focus (Tamamaki et al., 1995). Figure 13.4B shows the extrathalamic branches innervating the red nucleus and midbrain reticular nuclei, which give rise, respectively, to the rubrospinal and reticulospinal tracts. Figure 13.4B,C shows the extrathalamic branches innervating the midbrain tegmentum, which give rise to the tectospinal tract. Figure 13.4D shows the extrathalamic branches innervating multiple brainstem sites, many of which are sources of projections to spinal cord (red arrows), and the axon continues into the spinal cord itself.

Figure 13.5A shows an updated version of Figure 6.1 with the addition of the branching driver inputs to thalamus that target extrathalamic motor centers. While it appears that all layer 5 corticofugal axons branch repeatedly to innervate multiple targets, including many motor centers, not all branch to innervate thalamus as well. However, this pattern may be variable. On the one hand, Bourassa and Deschênes (1995) find that all eight of their labeled layer 5 efferents from visual cortex innervate multiple subcortical sites, of which six also innervate thalamus. Likewise, Economo et al. (2018) describe two different classes of layer 5 corticofugal cells in the anterior lateral motor cortex: one class branches to innervate thalamus and the other fails to innervate thalamus. On the other hand, Kita and Kita (2012) labeled 26 layer 5 corticofugal axons from motor cortex and found that every one had a branch innervating thalamus. While it may not yet be entirely clear the extent to which layer 5 corticofugal axons innervate thalamus, it does seem to be a common pattern, and every cortical area so far studied has a *population* of layer 5 corticofugal axons that innervate thalamus and extrathalamic targets in this fashion (Bourassa and Deschênes, 1995; Bourassa et al., 1995; Clasca et al., 1995; Economo et al., 2018; Prasad et al., 2020; Winnubst et al., 2019). Because of this branching pattern, the message sent to thalamus for relay to cortex is often or always a copy of a message sent to motor centers and thus can be considered an efference copy. Whether or not layer 5 axonal branches innervating motor centers play a modulatory or driving role in evoking postsynaptic responses remains to be determined. This is an important question to answer as it has important implications for the role these inputs play in the generation of motor commands, which, almost certainly, involves multiple circuits and sources of input.

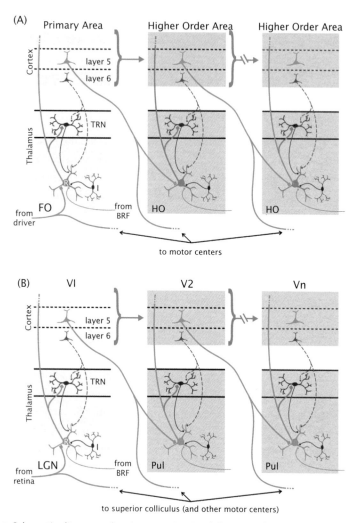

FIGURE 13.5 Schematic diagrams showing organizational features of first and higher order thalamic nuclei. A. A first order nucleus (*left*) represents the first relay of a particular type of subcortical information to a first order or primary cortical area. A higher order nucleus (*center and right*) relays information from layer 5 of one cortical area up the hierarchy to another cortical area. This relay can be from a primary area to a higher one (*center*) or between two higher order cortical areas (*right*). The important difference between first and higher order nuclei is the driver input, which is subcortical for a first order relay and from layer 5 of cortex for a higher order relay. Note that all thalamic nuclei receive an input from layer 6 of cortex, which is mostly organized in a reciprocal feedback manner, but higher order nuclei in addition receive a layer 5 input from cortex, which is feedforward. Note that the driver inputs, both subcortical and from layer 5, are typically from branching axons, with some extrathalamic targets being subcortical motor centers, and the significance of this is elaborated in the text. B. Example using the visual system as a model. The lateral geniculate nucleus is the first order thalamic nucleus, and the pulvinar is the higher order nucleus. The extrathalamic targets of the retinal input include the superior colliculus and other midbrain regions.

Abbreviations: BRF, brainstem reticular formation; *FO,* first order; *HO,* higher order *I,* interneuron; *R,* relay cell; *TRN,* thalamic reticular nucleus. Redrawn from Sherman, 2007.

13.2.1 Layer 5 Corticofugal Outputs Are the Efference Messages to Be Copied

Much of our behavior does not involve cortical control: think of chewing gum, most breathing, walking a familiar path, or chewing food. These behaviors are commonly controlled strictly by subcortical circuits. Only when behaviors require our attention and conscious effort does cortex come into play to dominate motor control. This is accomplished strictly via layer 5 corticofugal outputs. A point that cannot be overemphasized here is that, with all its beautiful circuitry and computational power, cortex would be pretty useless to us without its layer 5 outputs. It follows from the preceding discussion of the ubiquity and importance of efference copies that any time an efference signal is transmitted, it must be accompanied by an efference copy.

Regarding this last point, as noted above, there appears to be a class of layer 5 corticofugal neuron that innervates multiple subcortical motor sites but not thalamus (Bourassa and Deschênes, 1995; Economo et al., 2018). The innervation patterns of these neurons offer no clear route for an efference copy that can produce the needed forward model in a timely fashion. There are at least three possible reasons for this. First, these thalamus-avoiding projections may provide modulatory rather than driving inputs and thus would not carry a message requiring an efference copy. Second, it may be that there is an efference copy originating from the midbrain target of these layer 5 thalamus-avoiding axons that is relayed through thalamus (e.g., Sommer and Wurtz, 2004a,b); this would require an additional synaptic delay compared to the suggested route emanating from branches of layer 5 axons that innervate thalamus directly and raises the question as to whether this extra thalamic delay creates too long a latency for the needed forward model involved with efference copies. Third, these ideas about efference copies may simply be wrong.

Figure 13.5B schematically shows how this might work, using the visual system as an example. Consider how this circuitry might be involved in the control of eye movements via innervation of the superior colliculus as the system responds to a novel visual stimulus. Green axons represent thalamic driver inputs that carry a message that can be read as efference copies. The process starts with a new message activated by a new stimulus sent from the retina that can initiate an eye movement toward the location of the novel stimulus. The same message is also relayed through the lateral geniculate nucleus to V1. Recall that this single message can have multiple effects in cortex. Some V1 circuits will treat the message as a purely sensory signal and others as an efference copy of a message that may lead or contribute to an eye movement. As the new environmental information is analyzed by the circuitry of V1, a new message is sent out via layer 5. This message is sent to the superior colliculus, perhaps contributing to a more accurate eye placement or a new movement altogether, and a copy is relayed via the pulvinar to V2. Again, this copy of the output message can be treated by some V2 circuitry as further analysis of the visual scene by V1 and by other V2 circuitry as an efference copy. This process is repeated up the cortical hierarchy, and one result is that, at each level of the hierarchy, every cortical area has information about signals contributing to eye movements at all lower levels.

An anatomical study in the cat provides information that might be relevant to this scenario. Using orthograde tracing via an autoradiographic technique, the corticotectal projections of 25 different cortical areas, mostly visual, were traced and their laminar terminations within the superior colliculus noted (Harting et al., 1992). We can assume both that these projections emanate from layer 5, since layer 6 projects effectively only to thalamus, and also that many or most of the innervating axons branch to innervate thalamus as well. Generally, as one moves from superficial layers in the tectum to deeper ones, one generally moves from more sensory to more motor domains. That is, although the inputs to the superficial layers have access to the deeper layers, which are closely linked to motor outputs for eye and head movements (Ozen et al., 2000; Schiller and Tehovnik, 2001; Tehovnik et al., 2002; Helms et al., 2004), the axons terminating in the deeper layers are more directly linked to the motor outputs. One generalization that might be made from the anatomical pathway tracing data is that, as one ascends the visual cortical hierarchy, the terminations are deeper within the tectum (Harting et al., 1992) and thus closer to motor outputs (e.g., those controlling eye movements). Add to this the observation that retinotectal axons terminate most dorsally in the superior colliculus (Graybiel, 1975; Harting and Guillery, 1976; Basso and May, 2017) and thus terminate further from motor outputs than do cortical inputs. Thus, a sort of visual hierarchy exists from retina to V1 and up the hierarchy of visual cortical areas, and this is roughly reflected in the layering of their inputs to the superior colliculus. That is, retinal inputs are most dorsal and furthest from the motor outputs, and inputs higher in the hierarchy are closer to these motor outputs. Thus, the higher in the visual hierarchy, the more effective control the superior colliculus for creating behaviors such as head and eye movements.

This, of course, is quite speculative, but the conclusion can be considered in the context of Figure 13.5B. Retinotectal input can perhaps initiate a movement, but weakly and imprecisely, and a copy of this input is relayed via the lateral geniculate nucleus to V1. V1 circuitry acts on this information to improve the motor command, sending a new message to the superior colliculus and a copy of this message to V2 via the pulvinar. And so on. As a result, as visual information is more thoroughly analyzed at each higher hierarchical level, that level also contains information about any behaviors that are affected by lower levels.

An observation consistent with this scenario (but far from proving it!) was made in monkeys trained to pursue a moving target that suddenly appeared in their visual fields (Osborne et al., 2007). After a fixation to the target, each monkey pursued it with smooth eye movements, but an analysis of the smooth pursuit showed that accuracy was initially poor and improved to asymptotic values over the next 50–100 msec of pursuit. In the context of Figure 13.5B and the above-mentioned data of the layering of inputs in the superior colliculus (Harting et al., 1992), one might expect this result of smooth pursuit accuracy. That is, retinal input provides critical information needed for the initial eye movement, but its effect is relatively noisy because of the location of terminals and lack of computational analysis of the target, an analysis limited by retinal circuitry; improvement occurs over the next 50–100 msec because cortical processing in a hierarchical fashion provides better quality messages to the superior colliculus resulting in more accurate control of the visual pursuit.

Finally, note that Figure 13.5B also shows direct ascending connections between cortical areas. Perhaps these direct circuits underlie basic analysis of the visual scene, and the transthalamic circuits complement this analysis with updates to account for behavior via efference copies. This combination enables the brain to accurately determine what is happening in the visual world, a process that requires the ability to disambiguate self-generated actions from environmental events.

13.2.2 Other Views of the Anatomy of Efference Copies

Most discussions of actual circuitry regarding efference copies are frustratingly abstract as regards anatomical details. There are exceptions, such as experiments from the Wurtz laboratory describing an efference copy signal that travels from the superior colliculus to the medial dorsal nucleus of the thalamus to cortex (Sommer and Wurtz, 2004a,b; Wurtz and Sommer, 2004). This can be seen as complementary to ideas presented here for inputs to thalamus carrying a copy of signals sent to motor centers. However, more commonly, the suggestion is that a singular efference copy is created at the end of a lengthy sensorimotor circuit close to the final output to motoneurons. Examples can be found in von Holst and Mittelstaedt (1950, fig. 4) and Crapse and Sommer (2008a, fig. 1a).

For motor control initiated at the cortical level via layer 5 descending outputs, such a single efference copy at the site of motoneuronal control seems implausible. To be useful, an efference copy of a cortically initiated motor command has to affect further sensory processing *before* the next processing stage where the next sensory input (the "reafference") is compared to the expectation derived from the forward model based on the efference copy (von Holst and Mittelstaedt, 1950). Thus, the efference copy for a cortically induced motor command has to be fed back into cortex as soon as possible after the layer 5 corticofugal message is generated. Figure 13.6A shows why an efference copy created near the final output to motoneurons would be too late to be effective. By the time the cortical command makes it through brainstem and spinal circuitry to motoneurons, it might already be too late, but the efference copy would then have to make its way back up the spinal cord to thalamus through an as yet to be identified pathway before it can be used to create a forward model. Figure 13.6B, which shows the circuitry outlined here, gets the efference copy into cortical circuitry much more quickly.

There is a certain logic to these speculations. This starts with the obvious point that the result of cortical evolution has been to provide a more flexible behavioral repertoire to deal with a complex and changing ecosystem. This enhances the survival prospects of the organism. We can add two other features to this logical argument. First, the only anatomical means by which cortex can influence behavior is via layer 5 subcortical projections. Every cortical area so far studied for the feature has a layer 5 projection both to thalamus and extrathalamic motor structures (Sherman and Guillery, 2013; Prasad et al., 2020), and perhaps this is true even for every column within a cortical area. Second, given the need for an efference copy to be closely associated with every motor command, this must be true for the commands emanating from cortex. Since virtually all information reaching cortex must pass through thalamus, it follows that getting a copy of the message carried by layer 5 efferents back into cortex requires a thalamic relay. The most plausible (to us) circuitry to achieve this involves the branching driver inputs to thalamus, including the layer 5 inputs to higher order

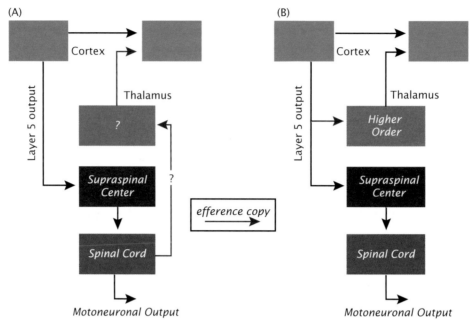

FIGURE 13.6 Two examples of possible efference copy pathways for motor commands originating in cortex. A. The efference copy is generated near the motor end of the command circuit, that is, in spinal cord near the recruited motoneurons. The problem with this scheme is getting the efference copy message back into the relevant cortical processing streams in terms of the relatively long delay and route through thalamus that has yet to be identified. B. Scheme offered by the present account, which provides a rapid delivery of the efference copy back into cortex through an identified thalamic route.

thalamus. While this hypothesis may or may not be appealing to the reader, it is worth emphasizing that the branching axons innervating thalamus are an anatomical fact, whereas the idea that this branching is related to efference copies is merely speculation, the point being that it is important ultimately to learn more about the functional significance of the branching axons.

13.3 An Evolutionary Perspective on Efference Copies

As noted above, efference copies must have appeared very early in the evolution of animals, and we can assume that our earliest vertebrate ancestors had such circuitry in their central nervous systems. Figure 13.3B shows a plausible example of an efference copy circuit associated with spinal circuitry that would likely have a very primitive ancestry. As evolution progressed from our mostly spinal ancestors, a variety of enhanced supraspinal circuits emerged to produce more flexible behaviors to allow organisms to adapt more effectively to their environmental challenges. Such evolution of additional efference circuitry would necessarily have to involve the co-evolution of matching efference copy circuitry. The evolutionary pinnacle of these supraspinal control circuits is the cerebral cortex, including requisite efference copy circuits.

13.3.1 Special Role of the Midbrain

As noted in Chapter 12.3.1, the evolution of cortex occurred without a co-evolution of a motor plant to which cortex has unique access. This means that, to influence behavior via its layer 5 outputs, cortex must operate through older circuitry in brainstem and spinal cord (Sherman and Usrey, 2021). It seems reasonable to assume that, in doing so, cortex would take advantage of circuits more advanced in evolutionary terms rather than having to "reinvent the wheel" by directly accessing the oldest circuitry in spinal cord and brainstem. In this regard, the most advanced sensorimotor center in nonmammalian vertebrates is the optic tectum and associated midbrain regions, which remains in mammals a major center for controlling head and body movements (Gaither and Stein, 1979; Stein and Gaither, 1983; Stein et al., 2009; Suzuki et al., 2019).

From this perspective, we suggest that the most efficient way for cortex to influence many or most behaviors is by operating through these brainstem structures. It is thus particularly interesting that every cortical area so far studied for this property exhibits a layer 5 projection to the midbrain (Deschênes et al., 1994; Bourassa and Deschênes, 1995; Bourassa et al., 1995; Kita and Kita, 2012; Economo et al., 2018; Prasad et al., 2020). Furthermore, evidence from these studies indicates that most or all layer 5 axons that innervate the midbrain branch to innervate thalamus. To the extent that the message sent to the midbrain by these layer 5 axons affects motor behavior, then that sent for relay through thalamus, being an exact copy of this message, is, by definition, an efference copy.

The idea is that the most common motor output of cortex operates via the midbrain with a copy of that message being relayed through thalamus to produce a forward model of the motor command for further cortical processing. However, there is at least one important proviso to this hypothesis. Economo et al. (2018) have shown for the anterior lateral motor cortex that many layer 5 axons in their branching fail to innervate thalamus, and so, if these do carry motor messages, there is no obvious route for an efference copy to be sent back to cortex. One possible explanation for this failure of layer 5 axons to innervate thalamus may lie in the functional role of the anterior lateral motor cortex; another may lie in the different functional role of these layer 5 projections. Clearly, there is still much to learn concerning the functional organization of cortical layer 5 outputs and their generality across cortex.

13.3.2 The Argument for Multiple Efference Copies Associated with Most Behaviors

Evolution is a messy process, and a general rule of the evolution of our brains is that older circuits that proved useful have been retained. As noted, much of our behavior does not involve cortex, and so these older subcortical circuits continue to control much behavior. However, the evolution of cortex did not occur with the co-evolution of a motor plant to which cortex has unique access. This means that cortex must compete and/or cooperate with these older brainstem and spinal circuits to affect behavior. It seems likely that most behaviors involving cortex involve the activation not just of layer 5 outputs but also of a complex combination of brainstem and spinal circuitry that these cortical outputs activate.

We suggest that the end result of this is that efference copies are generated for behaviors at multiple hierarchical levels that reflect evolutionary progress. The idea of efference copies at multiple levels has been covered previously by Crapse and Sommer (e.g., 2008a, fig. 1b, although these authors refer to these as "corollary discharges"). This implies that a given behavior involves not a single efference copy generated at the end of the sensorimotor chain, but multiple such copies at different levels. This, in turn, raises a question that has generally been ignored: How does the brain keep track of all of these efference copies, especially if some are not tied to motor commands that actually succeed in affecting behavior?

13.3.3 A Possible Reason for Transthalamic Circuitry

One of the questions about transthalamic circuitry raised in Chapter 6 is why any message transmitted between cortical areas should be routed through the thalamus. Figure 13.7A shows the actual circuit of these layer 5 outputs that innervate subcortical motor sites with a branch relayed through cortex to another cortical area. Figure 13.7B shows a hypothetical circuit in which the same information carried by the layer 5 branch is sent to the next

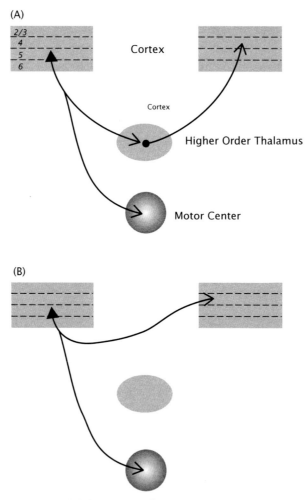

FIGURE 13.7 Non-thalamic and thalamic routes for cortically initiated efference copies. A. Thalamic route as suggested in the present account. B. Non-thalamic route. See text for details.

cortical area without a thalamic relay. Available evidence indicates that the circuit in Figure 13.7B does not exist, because the layer 5 cells that project to other cortical areas do not project subcortically, and those that do project subcortically do not project to other cortical areas (Petrof et al., 2012). The point is that evolution chose to route this information stream through thalamus (Figure 13.7A), thereby eschewing the direct route (Figure 13.7B). Why?

As noted above, it seems likely that the brain must keep track of multiple efference copies and account for the possibility that some motor commands do not lead to changes in behavior, meaning that any efference copies not associated with resultant behavior must be somehow invalidated. In the example of Figure 13.7B, if the message to motor centers from the afferent cortical area fails to affect behavior, the branch carrying the efference copy directly to the next cortical area cannot be stopped before arrival there. However, the example of Figure 13.7A affords the possibility of blocking the rogue efference copy at the level of the thalamus and preventing its further effect on cortical processing.

In this way, higher order thalamic relays could act as gates, which when closed could serve to block driver inputs, such as aberrant efference copies, from reaching cortex. As described in Chapter 6, it is interesting in this regard that higher order thalamic nuclei seem to have strong GABAergic inhibitory inputs that do not target first order nuclei, and such inputs could well serve to control thalamic gating, as suggested here.

13.4 Summary and Concluding Remarks

Figure 13.8 contrasts the conventional view of thalamocortical processing (Figure 13.8A) with the alternate view proposed here (Figure 13.8B). Figure 13.8A is the version generally provided in textbook accounts (Kandel et al., 2000, fig. 18.2, or Purves et al., 2012, fig. 12.18), and it can be summed up as follows (Felleman and Van Essen, 1991). Information first passes through thalamus to reach cortex, an example being the relay of retinal information to visual cortex via the lateral geniculate nucleus. This information is then processed through a hierarchical series of sensory, sensorimotor, and finally executive motor areas. From this executive level, a corticofugal message is sent to affect some behavior. Finally, at some abstractly considered position near the end of this circuit, an efference copy message is provided. This represents a sensorimotor circuit based on a chain of glutamatergic neurons that defines a functional input/output circuit involved in the transmission and processing of information leading to a behavioral result. Challenges to this view follow.

13.4.1 An Alternative to Concepts of "Sensory" and "Motor" Cortex

Note that the scheme of Figure 13.8A has no role for most of thalamus that is now recognized as higher order. That is, in the conventional scheme, once information is relayed to cortex, that information is processed strictly within cortex with no further role for thalamus until some final, executive stage is reached.

Note also that the conventional scheme of Figure 13.8A has single entry and exit points for cortical processing with much neuronal processing and time intervening. This seems implausible based on what we *think* we know about the brain evolution. That is, the notion that

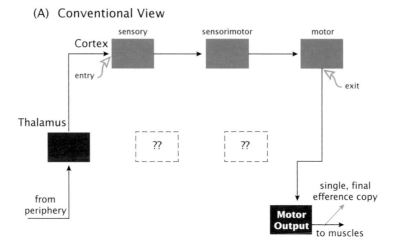

(A) Conventional View

(B) Alternative View

FIGURE 13.8 Comparison of conventional view (A) with the alternative view proposed here (B). The question marks in *A* indicate higher order thalamic relays, for which no specific function is suggested in this scheme. Note that in this conventional view (A), cortical processing in the sensorimotor circuit has a single entry and exit point. Further details in text.

Abbreviations: *FO*, first order; *HO*, higher order.

sensorimotor processing could involve so many steps (and so much time) as suggested by Figure 13.8A before a behavioral response ensues from any new and potentially threatening sensory message seems unlikely: any time a new sensory receptor or peripheral sensory process evolves, it will have no survival value if it lacks a fairly immediate motor output. In this sense, the scheme of Figure 13.8B seems more likely, with early sensory inputs having a relatively immediate connection to motor outputs.

Another important difference between the schemes is that Figure 13.8A shows a clear difference between sensory cortex, which receives new subcortical sensory information but has no subcortical (or motor) outputs, and motor cortex, which receives no new subcortical sensory information but does produce motor outputs. As Figure 13.8B suggests, every cortical area so far studied in this context has both input from thalamus as well as a layer 5 output that innervates subcortical motor structures (Prasad et al., 2020; reviewed in Sherman and Guillery, 2013). Thus, the common distinction between sensory and motor cortex seems misleading: the differences are more quantitative than qualitative. That is, what is considered "sensory" may be viewed as lower in the hierarchy of information processing, whereas "motor" areas are higher and benefit from more thorough processing of the environment. This reasoning predicts that "sensory" cortex does affect behavior with a shorter latency but with less accuracy and precision than do "motor" areas. Consistent with this notion is the anatomical study of Harting et al. (1992) and the eye tracking study of Osborne et al. (2007), both described in the section 13.2.1 above.

We thus suggest that the terminology of "sensory cortex" versus "motor cortex" be reconsidered. All cortical areas (and, indeed, all brain circuits) evolve to produce fairly immediate behavioral changes, and cortical areas vary in this regard quantitatively, mainly in hierarchical level and efficacy, rather than qualitatively.

13.4.2 Circuitry of Efference Copies

Finally, as we have argued in section 13.2.2 above, a single efference copy at the end of a sensorimotor circuit as suggested by Figure 13.8A simply seems wrong on the basis of timing requirements. Figure 13.8B shows what we regard as more plausible circuitry for getting efference copies in cortical processing by creating them at multiple hierarchical stages from the layer 5 motor outputs of every active cortical area.

13.5 Some Outstanding Questions

1. Do all driver inputs to thalamus branch to innervate extrathalamic targets, and, if some do not, how do they differ functionally from those that do?
2. If the single message carried by thalamocortical axons can be read by some targets as an efference copy and by others as other (e.g., sensory) information, what distinguishes these different targets?
3. Given that some descending layer 5 axons send a branch to the thalamus and some do not, what other functional characteristics distinguish these two types of axon arising from the same area of cortex?
4. What are the mechanisms that control the ability of layer 5 projections to dominate their subcortical motor targets as a consequence of attentional mechanisms? Are these outputs controlled at the cortical level (e.g., by gating their outputs) or at their targets?
5. Are there multiple efference copy messages created for complex behaviors, especially those initiated by cortex, and, if so, how does the brain handle them?

Conclusions

The central tenet to the ideas presented in this book is that neither the thalamus nor the cortex can be understood in any meaningful way in isolation of the other. Rather, proper function of cortex and thalamus requires thalamocortical interactions, mediated by thalamocortical and corticothalamic projections, each of which includes multiple pathways with specific anatomical and physiological properties. Thus, the traditional view that the thalamus does little more than pass the baton of sensory signals from periphery to cortex has been replaced with the modern view that the thalamus dynamically regulates the nature and flow of information to cortex as well as between cortical areas. Throughout the chapters in this book, we have intermixed generally accepted "facts" with "speculation," and we have tried to be clear when we transitioned between each of these. As stated in the Introduction, our primary goal of this book has not so much been to get the reader to accept our hypotheses and speculations, but rather to encourage skepticism and rethinking of standard textbook accounts of the subject. Some of the major themes emphasized throughout the book include: (1) the need for a proper classification of thalamocortical and corticothalamic circuits; (2) the role of spike timing for thalamocortical and corticothalamic communication and the mechanisms for modulating spike timing; (3) the organization and function of corticothalamic feedback projections; (4) the role of higher order thalamic nuclei in cortico-cortical communication and cortical functioning; (5) attentional modulation of thalamocortical interactions; and (6) a rethinking of efference copies and distinguishing neural signals as sensory versus motor.

14.1 Take-Away Points to Ponder

The following list highlights 10 key properties of thalamic processing and thalamocortical interactions that we have discussed throughout this book, along with their possible consequences.

- There is a spatial organization of glutamatergic input to thalamic relay cells. Feedforward input is generally directed to more proximal regions on the dendrites and involves ionotropic receptors, whereas feedback input from cortex is directed to more distal dendritic regions and involves both ionotropic and metabotropic receptors.

- As a consequence of this organization (and other factors), feedback input is proposed to modulate the excitatory state of relay cells and thereby their responsiveness to incoming driving input.

- Neurons in the thalamic reticular nucleus receive feedforward and feedback glutamatergic input from thalamocortical and corticothalamic axons, respectively. Thalamic reticular neurons, like other thalamic neurons, also receive nonglutamatergic input from a variety of nonthalamic sources.

- As a consequence, GABAergic input from thalamic reticular neurons onto relay cells is thought to be governed by behavioral state and the integration of feedforward and feedback activity levels, presumably to regulate communication between thalamus and cortex.

- Thalamic relay cells receive GABAergic inhibition from local interneurons via dendritic and axonal release of GABA and from neurons located in the thalamic reticular nucleus via axonal release.

- The spatial organization of these inputs, local inhibition onto the proximal dendrites of relay cells, and reticular inhibition onto more distal dendrites, suggest distinct contributions to response transience, gain control, and the burst/tonic state of relay cells.

- The time–voltage relationship for the de-inactivation of T-type Ca^{2+} channels determines whether thalamic neurons respond to membrane depolarization with burst spikes or tonic spikes.

- Because the membrane potential history is the variable that determines burst versus tonic spike firing, any input that hyperpolarizes a relay cell sufficiently in magnitude and time will shift the cell from tonic to burst firing mode; the opposite is also true in that any input that sufficiently depolarizes the cell will promote tonic firing.

- The timing (i.e., interspike interval) of arriving inputs, both to and from thalamus, determines their efficacy in evoking postsynaptic spikes.

- Based on this relationship and the nonlinearity that governs their production, burst spikes from the thalamus are particularly effective in driving cortical responses but less useful in conveying high-fidelity information. In contrast, tonic spikes follow sensory stimulation with greater fidelity.

- There are two distinct projection patterns from thalamus to cortex: one terminates topographically and primarily in middle layers, the other terminates diffusely and primarily in the upper layers, particularly layer 1.

- These projection patterns distinguish core versus matrix thalamus, respectively.
- Core projections are thought to bring specific and topographic information to cortex, whereas the matrix projections are thought to serve a diffuse role in adjusting cortical activity levels to reflect changes in behavioral state.

- A minority of the synapses made onto a thalamic neuron have a driving function; the majority are modulatory.

- The distinction of driver versus modulator input has been useful in classifying thalamic nuclei as first order or higher order. Relay cells in first order nuclei receive their driving input from subcortical sources, whereas relay cells in higher order nuclei receive their driving input from layer 5 corticothalamic axons.

- The source of the driving input to higher order thalamic nuclei, which comprises the majority of the thalamus, has helped to establish insight into function of these nuclei, which previously has been described using generic terms such as "association" thalamus.
- In contrast to first order relay cells that provide the cortex with information from subcortical sources, including sensory information from the periphery, higher order relay cells serve in the communication of cortical signals between cortical areas.

- There are two pathways from cortex to thalamus, one from layer 6 and the other from layer 5. All thalamic nuclei receive input from cortical layer 6, whereas a subset of nuclei also receive input from cortical layer 5.
  - The layer 6 projections serve to modulate both first order and higher order thalamic nuclei, whereas the layer 5 projections provide higher order thalamic nuclei with information that can be communicated to other cortical areas.
- Attention modulates the firing rate of thalamic neurons, the strength of thalamocortical communication, and thalamic contributions to interareal rhythmic activity between cortical areas.
  - The involvement of thalamus in attention suggests that multiple mechanisms underlie attention. This is an idea supported by attention networks in structures, such as the brainstem and midbrain, that predate the evolution of cortex.
- Neurons in layer 5 of cortex, with their axons to subcortical motor targets, are bottlenecks through which all cortical signals must pass to affect behavior. These axons often send a branch into higher order thalamic nuclei, such as the pulvinar.
  - The branching of layer 5 axons to target higher order thalamus in addition to motor targets may serve as an efferent copy, informing thalamus about signals being delivered to motor structures.

14.2 Why Do We Have a Thalamus?

Why have a thalamus, or, for a specific example, why do retinofugal axons not project directly to visual cortex? Olfactory information reaches cortex from the olfactory bulb without a thalamic relay, so it is a possible solution through evolution to process information at a cortical level without a thalamus. However, the cortex targeted in this example is paleocortex, not neocortex. Thalamus does seem to be a requisite relay station for neocortex. We can offer several explanations for the evolutionary creation of a thalamus.

14.2.1 Thalamus as a Bottleneck for Information Flow

Based on the take-away points listed above, a convincing argument can be made that the thalamus plays a critical role in controlling the delivery of signals to cortex from subcortical sites and between cortical areas, as well as in transforming the temporal structure of signals being relayed based on the integration of inputs from driving and modulatory networks.

While this broad role for thalamus is undoubtedly correct, a fair question to ask is what is gained by having the thalamus and thalamocortical circuits serve this function rather than being reliant on cortical circuits exclusively? One possible answer to this question is efficiency.

Thalamic nuclei are "bottlenecks" through which signals must pass to reach the cortex. Because the number of thalamic neurons (e.g., within the lateral geniculate nucleus) is orders of magnitude less than the number of neurons in the cortical areas to which they project, significantly fewer synapses are needed to perform the functions described above when performed in the thalamus compared to the hypothetical scenario of having them performed within the cortex. For instance, to shut down a message, as might occur during attention, requires far less neuronal machinery and energy if the message is intercepted at a thalamic level before reaching cortex. Although the number of feedforward-driving synapses to cortical target neurons may not be affected by not having a precortical bottleneck, the number of modulatory synapses would almost certainly be drastically increased. Within the thalamus, modulatory synapses outnumber driving synapses by approximately 20:1; thus, the number of additional modulatory synapses that would be needed to achieve similar functions within the cortex, where the number of neurons involved is much greater, would be extremely costly, in terms of both energy and space requirements.

14.2.2 Value of Burst and Tonic Firing

We suggested in Chapter 10 that burst firing serves as a "wake-up call" that strongly notifies cortex about a change in the environment. If true, this would be another obvious advantage to having a thalamic relay that can use the burst mode of its relay cells to emphasize that a new message is being relayed. This feature could not plausibly be incorporated if, for instance, the retina projected directly to cortex.

14.2.3 Thalamus as an Essential Player in Corticocortical Cooperation

Perhaps transthalamic pathways provide the means whereby different cortical areas cooperate for various cognitive functions, including attention (Seidemann et al., 1998; Andersen and Cui, 2009; Fries, 2009; Saalmann et al., 2012; Halassa and Kastner, 2017). For instance, recent evidence using optogenetics and Ca^{2+} imaging in mice indicates that activation of cortical layer 5 cells produces waves of activity in other cortical areas that depend largely on transthalamic pathways (Stroh et al., 2013). As noted in Chapter 6, a common feature of transthalamic pathways is that they are often if not always paralleled by direct connections between the same cortical areas. A given cortical area generally connects directly with multiple other cortical areas, and, if these are connected also via transthalamic pathways, it may be that those areas connected by both pathways, being active, form working, cooperative groupings in a sort of AND gate configuration. It is then the gating by higher order thalamus that determines the pattern of cortical areas joined in a cooperative alliance (Seidemann et al., 1998; Andersen and Cui, 2009; Fries, 2009; Saalmann et al., 2012; Halassa and Kastner, 2017).

14.2.4 Thalamus Is a Mysterious Product of a Chaotic Evolutionary Process

We must admit that we really cannot explain in detail how evolution produced so many features of our bodies. It operates like a pseudorandom hit-or-miss process, discarding the misses that fail along the way. The final process might have a complex history that becomes impossible to reconstruct. The point is that an intelligent designer might produce a very different way to link neocortex to the rest of the brain without a thalamus. In other words, our suggestions for why we have a thalamus might be read as justification after the fact. Evolution produced the thalamus, but how and why remain speculation only.

14.3 Summary and Concluding Remarks

Finally, throughout this book we have ended each chapter with a set of unanswered questions. The purpose of these questions was not simply to state what we do not know, but rather to stimulate further thinking and, hopefully, plant a few seeds for the generation of new and testable ideas that will further advance our understanding of thalamocortical interactions and their critical role in brain function. As we think about these questions and possible approaches for finding answers, the importance of studying thalamocortical interactions across systems and in a diversity of species cannot be stressed too strongly. Each system and species provides unique "toeholds" from which mechanism and function can be explored and understood. As common strategies are identified, these strategies can be viewed as general rules, and exceptions to these rules are important, because they represent alternative approaches for success that may be equally or more effective in challenging settings and different environments.

References

Adams MM, Hof PR, Gattass R, Webster MJ, Ungerleider LG (2000) Visual cortical projections and chemoarchitecture of macaque monkey pulvinar. J Comp Neurol 419:377–393.

Ahissar E, Sosnik R, Haidarliu S (2000) Transformation from temporal to rate coding in a somatosensory thalamocortical pathway. Nature 406:302–306.

Ahmed B, Anderson JC, Martin KAC, Nelson JC (1997) Map of the synapses onto layer 4 basket cells of the primary visual cortex of the cat. J Comp Neurol 380:230–242.

Alitto HJ, Rathbun DL, Vandeleest JJ, Alexander PC, Usrey WM (2019) The augmentation of retinogeniculate communication during thalamic burst mode. J Neurosci 39:5710.

Alitto HJ, Moore BD, Rathbun DL, Usrey WM (2011) A comparison of visual responses in the lateral geniculate nucleus of alert and anaesthetized macaque monkeys. J Physiol 589(Pt 1):87–99.

Alitto HJ, Usrey WM (2003) Corticothalamic feedback and sensory processing. Curr Opin Neurobiol 13:440–445.

Alitto HJ, Usrey WM (2004) Influence of contrast on orientation and temporal frequency tuning in ferret primary visual cortex. J Neurophysiol 91:2797–2808.

Alitto HJ, Usrey WM (2008) Origin and dynamics of extraclassical suppression in the lateral geniculate nucleus of the macaque monkey. Neuron 57:135–146.

Alitto HJ, Weyand TG, Usrey WM (2005) Distinct properties of stimulus-evoked bursts in the lateral geniculate nucleus. J Neurosci 25:514–523.

Alonso JM, Martinez LM (1998) Functional connectivity between simple cells and complex cells in cat striate cortex. Nat Neurosci 1:395–403.

Alonso JM, Usrey WM, Reid RC (1996) Precisely correlated firing in cells of the lateral geniculate nucleus. Nature 383:815–819.

Alonso JM, Usrey WM, Reid RC (2001) Rules of connectivity between geniculate cells and simple cells in cat primary visual cortex. J Neurosci 21:4002–4015.

Amadeo A, Ortino B, Frassoni C (2001) Parvalbumin and GABA in the developing somatosensory thalamus of the rat: An immunocytochemical ultrastructural correlation. Anat Embryol (Berl) 203:109–119.

Andersen RA, Cui H (2009) Intention, action planning, and decision making in parietal-frontal circuits. Neuron 63:568–583.

Anwyl R (2009) Metabotropic glutamate receptor-dependent long-term potentiation. Neuropharmacol 56:735–740.

Applebury ML, Antoch MP, Baxter LC, Chun LL, Falk JD, Farhangfar F, Kage K, Krzystolik MG, Lyass LA, Robbins JT (2000) The murine cone photoreceptor: A single cone type expresses both S and M opsins with retinal spatial patterning. Neuron 27(3):513–523.

Arcaro MJ, Pinsk MA, Chen J, Kastner S (2018) Organizing principles of pulvino-cortical functional coupling in humans. Nat Commun 9(1):5382 doi:10.1038/s41467-018-07725-6.

Arcaro MJ, Pinsk MA, Kastner S (2015) The anatomical and functional organization of the human visual pulvinar. J Neurosci 35:9848–9871.

Arcelli P, Frassoni C, Regondi MC, De Biasi S, Spreafico R (1997) GABAergic neurons in mammalian thalamus: A marker of thalamic complexity? Brain Res Bull 42:27–37.

Archer DR, Alitto HJ, Usrey WM. (2021) Stimulus contrast affects spatial integration in the lateral geniculate nucleus of macaque monkeys. J Neurosci. 41:6246–6256.

Awh E, Belopolsky AV, Theeuwes J (2012) Top-down versus bottom-up attentional control: A failed theoretical dichotomy. Trends Cogn Sci 16(8):437–443.

Azouz R, Gray CM (2000) Dynamic spike threshold reveals a mechanism for synaptic coincidence detection in cortical neurons in vivo. Proc Natl Acad Sci USA 97(14):8110–8115.

Baden T, Berens P, Franke K, Roman RM, Bethge M, Euler T (2016) The functional diversity of retinal ganglion cells in the mouse. Nature 529(7586):345–350.

Bajo VM, Nodal FR, Bizley JK, Moore DR, King AJ (2007) The ferret auditory cortex: Descending projections to the inferior colliculus. Cereb Cortex 17:475–491.

Baldauf ZB, Chomsung RD, Carden WB, May PJ, Bickford ME (2005) Ultrastructural analysis of projections to the pulvinar nucleus of the cat. I: Middle suprasylvian gyrus (areas 5 and 7). J Comp Neurol 485:87–107.

Baldwin MK, Balaram P, Kaas JH (2013) Projections of the superior colliculus to the pulvinar in prosimian galagos (Otolemur garnettii) and VGLUT2 staining of the visual pulvinar. J Comp Neurol 521:1664–1682.

Baldwin MKL, Balaram P, Kaas JH (2017) The evolution and functions of nuclei of the visual pulvinar in primates. J Comp Neurol 525:3207–3226.

Banerjee A, Larsen RS, Philpot BD, Paulsen O (2016) Roles of presynaptic NMDA receptors in neurotransmission and plasticity. Trends Neurosci 39(1):26–39.

Barlow HB, Hill RM, Levick WR (1964) Retinal ganglion cells responding selectively to direction and speed of image motion in the rabbit. J Physiol 173:377–407.

Barre A, Berthoux C, De BD, Valjent E, Bockaert J, Marin P, Becamel C (2016) Presynaptic serotonin 2A receptors modulate thalamocortical plasticity and associative learning. Proc Natl Acad Sci USA 113(10):E1382–E1391.

Barthó P, Freund TF, Acsády L (2002) Selective GABAergic innervation of thalamic nuclei from zona incerta. Eur J Neurosci 16:999–1014.

Basso MA, Bickford ME, Cang J (2021) Unraveling circuits of visual perception and cognition through the superior colliculus. Neuron 109:918–937.

Basso MA, May PJ (2017) Circuits for action and cognition: A view from the superior colliculus. Annu Rev Vis Sci 3:197–226.

Bastos AM, Briggs F, Alitto HJ, Mangun GR, Usrey WM (2014) Simultaneous recordings from the primary visual cortex and lateral geniculate nucleus reveal rhythmic interactions and a cortical source for gamma-band oscillations. J Neurosci 34:7639–7644.

Bedard C, Destexhe A (2016) Generalized cable models of neurons and dendrites. In: Neuroscience in the 21th century (Pfaff DW, Volkow ND, eds), pp 1–11. New York: Springer.

Beierlein M (2014) Synaptic mechanisms underlying cholinergic control of thalamic reticular nucleus neurons. J Physiol 592(19):4137–4145.

Beltramo R, Scanziani M (2019) A collicular visual cortex: Neocortical space for an ancient midbrain visual structure. Science 363:64–69.

Bender DB (1983) Visual activation of neurons in the primate pulvinar depends on cortex but not colliculus. Brain Res 279:258–261.

Benson DL, Isackson PJ, Hendry SH, Jones EG (1991) Differential gene expression for glutamic acid decarboxylase and type II calcium-calmodulin-dependent protein kinase in basal ganglia, thalamus, and hypothalamus of the monkey. J Neurosci 11:1540–1564.

Berkley KJ (1975) Different targets of different neurons in nucleus gracilis of the cat. J Comp Neurol 163:285–303.

Berkley KJ, Blomqvist A, Pelt A, Fink R (1980) Differences in the collateralization of neural projections from dorsal column nuclei and lateral cervical nucleus to the thalamus and tectum in the cat: An anatomical study using two different double labeling techniques. Brain Res 202:273–290.

Berman NEJ, Jones EG (1977) A retino-pulvinar projection in the cat. Brain Res 134:237–248.

Berson DM, Isayama T, Pu M (1999) The eta ganglion cell type of cat retina. J Comp Neurol 408:204–219.

Berson DM, Pu M, Famiglietti EV (1998) The zeta cell: A new ganglion cell type in cat retina. J Comp Neurol 399:269–288.

Bezdudnaya T, Cano M, Bereshpolova Y, Stoelzel CR, Alonso JM, Swadlow HA (2006) Thalamic burst mode and inattention in the awake LGNd. Neuron 49:421–432.

Bickford ME, Zhou N, Krahe TE, Govindaiah G, Guido W (2015) Retinal and tectal "driver-like" inputs converge in the shell of the mouse dorsal lateral geniculate nucleus. J Neurosci 35:10523–10534.

Blasdel GG, Lund JS (1983) Termination of afferent axons in macaque striate cortex. J Neurosci 3:1389–1413.

Bloomfield SA, Sherman SM (1989) Dendritic current flow in relay cells and interneurons of the cat's lateral geniculate nucleus. Proc Natl Acad Sci USA 86:3911–3914.

Blot A, Roth MM, Gasler IT, Javadzadeh M, Imhof F, Hofer SB (2021) Visual intracortical and transthalamic pathways carry distinct information to cortical areas. Neuron 109:1996–2008.

Bokor H, Frere SGA, Eyre MD, Slezia A, Ulbert I, Luthi A, Acsády L (2005) Selective GABAergic control of higher-order thalamic relays. Neuron 45:929–940.

Boudreau CE, Ferster D (2005) Short-term depression in thalamocortical synapses of cat primary visual cortex. J Neurosci 25:7179–7190.

Bourassa J, Deschênes M (1995) Corticothalamic projections from the primary visual cortex in rats: A single fiber study using biocytin as an anterograde tracer. Neurosci 66:253–263.

Bourassa J, Pinault D, Deschênes M (1995) Corticothalamic projections from the cortical barrel field to the somatosensory thalamus in rats: A single-fibre study using biocytin as an anterograde tracer. Eur J Neurosci 7:19–30.

Boycott BB, Wässle H (1974) The morphological types of ganglion cells of the domestic cat's retina. J Physiol (Lond) 240:397–419.

Branco T, Hausser M (2011) Synaptic integration gradients in single cortical pyramidal cell dendrites. Neuron 69(5):885–892.

Brasier DJ, Feldman DE (2008) Synapse-specific expression of functional presynaptic NMDA receptors in rat somatosensory cortex. J Neurosci 28(9):2199–2211.

Brickman AM, Buchsbaum MS, Shihabuddin L, Byne W, Newmark RE, Brand J, Ahmed S, Mitelman SA, Hazlett EA (2004) Thalamus size and outcome in schizophrenia. Schizophr Res 71:473–484.

Briggs F, Kiley CW, Callaway EM, Usrey WM (2016) Morphological substrates for parallel streams of corticogeniculate feedback originating in both V1 and V2 of the macaque monkey. Neuron 90:388–399.

Briggs F, Mangun GR, Usrey WM (2013) Attention enhances synaptic efficacy and the signal-to-noise ratio in neural circuits. Nature 499:476–480.

Briggs F, Usrey WM (2005) Temporal properties of feedforward and feedback pathways between the thalamus and visual cortex in the ferret. Thalamus Relat Syst 3:133–139.

Briggs F, Usrey WM (2008) Emerging views of corticothalamic function. Curr Opin Neurobiol 18:403–407.

Briggs F, Usrey WM (2009) Parallel processing in the corticogeniculate pathway of the macaque monkey. Neuron 62:135–146.

Briggs F, Usrey WM (2011) Corticogeniculate feedback and visual processing in the primate. J Physiol 589(Pt 1):33–40.

Briggs F, Usrey WM (2014) Functional properties of cortical feedback to the primate lateral geniculate nucleus. In: The new visual neurosciences (Chalupa LM, Werner JS, eds), pp 315–322. Cambridge, MA: MIT Press.

Brodmann K (1909) Vergleichende Lokalisationslehre der Grosshirnrinde in ihren Prinzipien dargestellt auf Grund des Zeelenbaues. Leipzig: Johann Ambrosius Barth Verlag.

Brody CD (1998) Slow covariations in neuronal resting potentials can lead to artefactually fast cross-correlations in their spike trains. J Neurophysiol 80:3345–3351.

Brown SP, Mathur BN, Olsen SR, Luppi PH, Bickford ME, Citri A (2017) New breakthroughs in understanding the role of functional interactions between the neocortex and the claustrum. J Neurosci 37:10877–10881.

Bruno RM, Sakmann B (2006) Cortex is driven by weak but synchronously active thalamocortical synapses. Science 312:1622–1627.

Bull MS, Berkley KJ (1984) Differences in the neurones that project from the dorsal column nuclei to the diencephalon, pretectum, and the tectum in the cat. Somatosens Res 1:281–300.

Bunt AH, Hendrickson AE, Lund JS, Lund RD, Fuchs AF (1975) Monkey retinal ganglion cells: Morphometric analysis and tracing of axonal projections, with a consideration of the peroxidase technique. J Comp Neurol 164:265–285.

Buschman TJ, Kastner S (2015) From behavior to neural dynamics: An integrated theory of attention. Neuron 88(1):127–144.

Butts DA, Cui Y, Casti AR (2016) Nonlinear computations shaping temporal processing of precortical vision. J Neurophysiol 116:1344–1357.

Butts DA, Weng C, Jin J, Alonso JM, Paninski L (2011) Temporal precision in the visual pathway through the interplay of excitation and stimulus-driven suppression. J Neurosci 31:11313–11327.

Byne W, Hazlett EA, Buchsbaum MS, Kemether E (2009) The thalamus and schizophrenia: Current status of research. Acta Neuropathol 117:347–368.

Cajal SRy (1899a) Estudios sobre la corteza cerebral humanan I: Corteza visual. Revista Trimestral Micrografica 4:1–63.

Cajal SRy (1899b) Estudios sobre la corteza cerebral humanan II: Estructura de la corteza motriz del hombre y mamiferos superiores. Revista Trimestral Micrografica 4:117–200.

Cajal SRy (1911) Histologie du Système Nerveaux de l'Homme et des Vertébrés. Paris: Maloine.

Cajal SRy (1933)?Neuronismo o Reticularismo? Las pruebas objetivas de la unidad anatomica de las celulas nerviosas. Arch Neurobiol Madrid 13:1–144.

Campbell PW, Govindaiah G, Masterson SP, Bickford ME, Guido W (2020) Synaptic properties of the feedback connections from the thalamic reticular nucleus to the dorsal lateral geniculate nucleus. J Neurophysiol . 124:404–417.

Cappe C, Morel A, Rouiller EM (2007) Thalamocortical and the dual pattern of corticothalamic projections of the posterior parietal cortex in macaque monkeys. Neurosci 146:1371–1387.

Caputi A, Rozov A, Blatow M, Monyer H (2009) Two calretinin-positive GABAergic cell types in layer 2/3 of the mouse neocortex provide different forms of inhibition. Cereb Cortex 19:1345–1359.

Carandini M, Ferster D (1997) A tonic hyperpolarization underlying contrast adaptation in cat visual cortex. Science 276:949–952.

Carandini M, Horton JC, Sincich LC (2007) Thalamic filtering of retinal spike trains by postsynaptic summation. J Vis 7(14):1–11.

Carden WB, Bickford ME (2002) Synaptic inputs of class III and class V interneurons in the cat pulvinar nucleus: Differential integration of RS and RL inputs. Visual Neurosci 19:51–59.

Carpenter MB, Nakano K, Kim R (1976) Nigrothalamic projections in the monkey demonstrated by autoradiographic technics. J Comp Neurol 165:401–415.

Carvalho LS, Pessoa DMA, Mountford JK, Davies WIL, Hunt DM (2017) The genetic and evolutionary drives behind primate color vision. Front Ecology Evolution 5:34.

Casagrande VA (1994) A third parallel visual pathway to primate area V1. Trends Neurosci 17:305–309.

Casagrande VA, Kaas JH (1994) The afferent, intrinsic, and efferent connections of primary visual cortex in primates. In: Cerbral cortex (Peters A, Rockland KS, eds), pp 201–259. New York: Plenum.

Casagrande VA, Xu X (2004) Parallel visual pathways: A comparative perspective. In: The visual neurosciences (Chalupa LM, Werner JS, eds), pp 494–506. Cambridge, MA: MIT Press.

Casagrande VA, Yazar F, Jones KD, Ding Y (2007) The mophology of the koniocellular axon pathway in the macaque monkey. Cereb Cortex 17:2334–2345.

Casanova C (1993) Response properties of neurons in area 17 projecting to the striate-recipient zone of the cat's lateralis posterior- pulvinar complex: Comparison with cortico-tectal cells. Exp Brain Res 96:247–259.

Catania KC (2011) The sense of touch in the star-nosed mole: From mechanoreceptors to the brain. Philos Trans R Soc Lond B Biol Sci 366(1581):3016–3025.

Catterall WA (2010) Ion channel voltage sensors: Structure, function, and pathophysiology. Neuron 67:915–928.

Catterall WA (2011) Voltage-gated calcium channels. Cold Spring Harb Perspect Biol 3:a003947.

Chalk M, Herrero JL, Gieselmann MA, Delicato LS, Gotthardt S, Thiele A (2010) Attention reduces stimulus-driven gamma frequency oscillations and spike field coherence in V1. Neuron 66:114–125.

Chalupa LM, Anchel H, Lindsley DB (1972) Visual input to the pulvinar via lateral geniculate, superior colliculus and visual cortex in the cat. Exp Neurol 36:449–462.

Chalupa LM, Thompson ID (1980) Retinal ganglion cell projections to the superior colliculus of the hamster demonstrated by the horseradish peroxidase technique. Neurosci Lett 19:13–19.

Chance FS, Abbott LF, Reyes A (2002) Gain modulation from background synaptic input. Neuron 35:773–782.

Chauveau F, Celerier A, Ognard R, Pierard C, Beracochea D (2005) Effects of ibotenic acid lesions of the mediodorsal thalamus on memory: Relationship with emotional processes in mice. Behav Brain Res 156:215–223.

Chen C, Blitz DM, Regehr WG (2002) Contributions of receptor desensitization and saturation to plasticity at the retinogeniculate synapse. Neuron 33:779–788.

Chen Y, Martinez-Conde S, Macknik SL, Bereshpolova Y, Swadlow HA, Alonso JM (2008) Task difficulty modulates the activity of specific neuronal populations in primary visual cortex. Nat Neurosci 11:974–982.

Cheong SK, Tailby C, Solomon SG, Martin PR (2013) Cortical-like receptive fields in the lateral geniculate nucleus of marmoset monkeys. J Neurosci 33:6864–6876.

Chomsung RD, Petry HM, Bickford ME (2008) Ultrastructural examination of diffuse and specific tectopulvinar projections in the tree shrew. J Comp Neurol 510:24–46.

Chung S, Li X, Nelson SB (2002) Short-term depression at thalamocortical synapses contributes to rapid adaptation of cortical sensory responses in vivo. Neuron 34:437–446.

Clarke E, O'Malley CD (1968) The human brain and the spinal cord. A historical study illustrated by writings from antiquity to the twentieth century. Berkeley: University of California Press.

Clasca F, Angelucci A, Sur M (1995) Layer-specific programs of development in neocortical projection neurons. Proc Natl Acad Sci USA 92:11145–11149.

Clasca F, Rubio-Garrido P, Jabaudon D (2012) Unveiling the diversity of thalamocortical neuron subtypes. Eur J Neurosci 35:1524–1532.

Cleland BG (1986) The dorsal lateral geniculate nucleus of the cat. In: Visual neuroscience (Pettigrew JD, Sanderson KJ, Levick WR, eds), pp 111–120. London: Cambridge University Press.

Cleland BG, Dubin MW, Levick WR (1971a) Simultaneous recording of input and output of lateral geniculate neurones. Nat New Biol 231:191–192.

Cleland BG, Dubin MW, Levick WR (1971b) Sustained and transient neurones in the cat's retina and lateral geniculate nucleus. J Physiol (Lond) 217:473–496.

Cleland BG, Levick WR (1974) Properties of rarely encountered types of ganglion cells in the cat's retina and an overall classification. J Physiol (Lond) 240:457–492.

Cleland BG, Levick WR, Morstyn R, Wagner HG (1976) Lateral geniculate relay of slowly conducting retinal afferents to cat visual cortex. J Physiol (Lond) 255:299–320.

Clemente-Perez A, Makinson SR, Higashikubo B, Brovarney S, Cho FS, Urry A, Holden SS, Wimer M, Dávid C, Fenno LE, Acsády L, Deisseroth K, Paz JT (2017) Distinct thalamic reticular cell types differentially modulate normal and pathological cortical rhythms. Cell Rep 19(10):2130–2142.

Cohen MR, Maunsell JH (2009) Attention improves performance primarily by reducing interneuronal correlations. Nat Neurosci 12:1594–1600.

Colonnier M, Guillery RW (1964) Synaptic organization in the lateral geniculate nucleus of the monkey. Z Zellforsch 62:333–355.

Conn PJ, Pin JP (1997) Pharmacology and functions of metabotropic glutamate receptors. Annu Rev Pharmacol Toxicol 37:205–237.

Contreras D, Steriade M (1995) Cellular basis of EEG slow rhythms: A study of dynamic corticothalamic relationships. J Neurosci 15:604–622.

Coogan TA, Burkhalter A (1990) Conserved patterns of cortico-cortical connections define areal hierarchy in rat visual cortex. Exp Brain Res 80:49–53.

Coogan TA, Burkhalter A (1993) Hierarchical organization of areas in rat visual cortex. J Neurosci 13:3749–3772.

Coomes DL, Schofield RM, Schofield BR (2005) Unilateral and bilateral projections from cortical cells to the inferior colliculus in guinea pigs. Brain Res 1042(1):62–72.

Coutinho V, Knopfel T (2002) Metabotropic glutamate receptors: Electrical and chemical signaling properties. Neuroscientist 8:551–561.

Covic EN, Sherman SM (2011) Synaptic properties of connections between the primary and secondary auditory cortices in mice. Cereb Cortex 21:2425–2441.

Cox CL, Denk W, Tank DW, Svoboda K (2000) Action potentials reliably invade axonal arbors of rat neocortical neurons. Proc Natl Acad Sci USA 97:9724–9728.

Cox CL, Sherman SM (2000) Control of dendritic outputs of inhibitory interneurons in the lateral geniculate nucleus. Neuron 27:597–610.

Cox CL, Zhou Q, Sherman SM (1998) Glutamate locally activates dendritic outputs of thalamic interneurons. Nature 394:478–482.

Crandall SR, Cruikshank SJ, Connors BW (2015) A corticothalamic switch: Controlling the thalamus with dynamic synapses. Neuron 86:768–782.

Crapse TB, Sommer MA (2008a) Corollary discharge across the animal kingdom. Nat Rev Neurosci 9:587–600.

Crapse TB, Sommer MA (2008b) Corollary discharge circuits in the primate brain. Curr Opin Neurobiol 18:552–557.

Crick F (1984) Function of the thalamic reticular complex: The searchlight hypothesis. Proc Natl Acad Sci USA 81:4586–4590.

Cronenwett WJ, Csernansky J (2010) Thalamic pathology in schizophrenia. Curr Top Behav Neurosci 4:509–528.

Croner LJ, Kaplan E (1995) Receptive fields of P and M ganglion cells across the primate retina. Vision Res 35:7–24.

Crook JD, Davenport CM, Peterson BB, Packer OS, Detwiler PB, Dacey DM (2009) Parallel ON and OFF cone bipolar inputs establish spatially coextensive receptive field structure of blue-yellow ganglion cells in primate retina. J Neurosci 29(26):8372–8387.

Crook JD, Packer OS, Dacey DM (2014a) A synaptic signature for ON- and OFF-center parasol ganglion cells of the primate retina. Vis Neurosci 31(1):57–84.

Crook JD, Packer OS, Troy JB, Dacey DM (2014b) Synaptic mechanisms of color and luminance coding: Rediscovering the X-Y dichotomy in primate retinal ganglion cells. In: The new visual neurosciences (Chalupa LM, Werner JS, eds), pp 257–283. Cambridge, MA: MIT Press.

Crook JD, Peterson BB, Packer OS, Robinson FR, Troy JB, Dacey DM (2008) Y-Cell receptive field and collicular projection of parasol ganglion cells in macaque monkey retina. J Neurosci 28:11277–11291.

Cusick CG, Gould HJ, III (1990) Connections between area 3b of the somatosensory cortex and subdivisions of the ventroposterior nuclear complex and the anterior pulvinar nucleus in squirrel monkeys. J Comp Neurol 292(1):83–102.

D'Souza RD, Burkhalter A (2017) A aminar organization for selective cortico-cortical communication. Front Neuroanat 11:71.

Dacey DM (1993) Morphology of a small-field bistratified ganglion cell type in the macaque and human retina. Visual Neurosci 10:1081–1098.

Dacey DM (1999) Primate retina: Cell types, circuits and color opponency. Prog Retin Eye Res 18(6):737–763.

Dacey DM (2000) Parallel pathways for spectral coding in primate retina. Annu Rev Neurosci 23:743–775.

Dacey DM, Lee BB (1994) The "blue-on" opponent pathway in primate retina originates from a distinct bistratified ganglion cell type. Nature 367:731–735.

Dacey DM, Packer OS (2003) Colour coding in the primate retina: Diverse cell types and cone-specific circuitry. Curr Opin Neurobiol 13(4):421–427.

Dacey DM, Peterson BB, Robinson FR, Gamlin PD (2003) Fireworks in the primate retina: In vitro photodynamics reveals diverse LGN-projecting ganglion cell types. Neuron 37:15–27.

Dan Y, Alonso JM, Usrey WM, Reid RC (1998) Coding of visual information by precisely correlated spikes in the lateral geniculate nucleus. Nat Neurosci 1:501–507.

Danos P, Baumann B, Kramer A, Bernstein HG, Stauch R, Krell D, Falkai P, Bogerts B (2003) Volumes of association thalamic nuclei in schizophrenia: A postmortem study. Schizophr Res 60:141–155.

Datskovskaia A, Carden WB, Bickford ME (2001) Y retinal terminals contact interneurons in the cat dorsal lateral geniculate nucleus. J Comp Neurol 430:85–100.

Datta S, Siwek DF (2002) Single cell activity patterns of pedunculopontine tegmentum neurons across the sleep-wake cycle in the freely moving rats. J Neurosci Res 70:611–621.

Dawson TM, Dawson VL (1996) Nitric oxide synthase: Role as a transmitter/mediator in the brain and endocrine system. Annu Rev Med 47:219–227.

De Monasterio FM (1978) Properties of concentrically organized X and Y ganglion cells of macaque retina. J Neurophysiol 41:1394–1417.

De Weerd P, Peralta MR, III, Desimone R, Ungerleider LG (1999) Loss of attentional stimulus selection after extrastriate cortical lesions in macaques. Nat Neurosci 2:753–758.

DeFelipe J, Jones EG (1988) Cajal on the cerebral cortex: An annotated translation of the complete writings. Oxford: Oxford University Press.

Deleuze C, David F, Behuret S, Sadoc G, Shin HS, Uebele VN, Renger JJ, Lambert RC, Leresche N, Bal T (2012) T-type calcium channels consolidate tonic action potential output of thalamic neurons to neocortex. J Neurosci 32:12228–12236.

Denman DJ, Contreras D (2016) On parallel streams through the mouse dorsal lateral geniculate nucleus. Front Neural Circuits 10:20.

Denning KS, Reinagel P (2005) Visual control of burst priming in the anesthetized lateral geniculate nucleus. J Neurosci 25:3531–3538.

DePasquale R, Sherman SM (2011) Synaptic properties of corticocortical connections between the primary and secondary visual cortical areas in the mouse. J Neurosci 31:16494–16506.

DePasquale R, Sherman SM (2012) Modulatory effects of metabotropic glutamate receptors on local cortical circuits. J Neurosci 32:7364–7372.

DePasquale R, Sherman SM (2013) A modulatory effect of the feedback from higher visual areas to V1 in the mouse. J Neurophysiol 109:2618–2631.

Derrington AM, Lennie P (1984) Spatial and temporal contrast sensitivities of neurones in lateral geniculate nucleus of macaque. J Physiol (Lond) 357:219–240.

Deschênes M, Bourassa J, Pinault D (1994) Corticothalamic projections from layer V cells in rat are collaterals of long-range corticofugal axons. Brain Res 664:215–219.

Deschênes M, Veinante P, Zhang ZW (1998) The organization of corticothalamic projections: Reciprocity versus parity. Brain Res Rev 28:286–308.

Desimone R, Duncan J (1995) Neural mechanisms of selective visual attention. Annu Rev Neurosci 18:193–222.

Devi LA, Fricker LD (2013) Transmitters and peptides: Basic principles. In: Neuroscience in the 21st century: From basic to clinical (Pfaff DW, ed), pp 1487–1503. New York: Springer .

Diamond IT, Conley M, Fitzpatrick D, Raczkowski D (1991) Evidence for separate pathways within the tecto-geniculate projection in the tree shrew. Proc Natl Acad Sci USA 88:1315–1319.

Diamond IT, Conley M, Itoh K, Fitzpatrick D (1985) Laminar organization of geniculocortical projections in *galago senegalensis* and *aotus trivirgatus*. J Comp Neurol 242:584–610.

Diamond ME, Armstrong-James M, Ebner FF (1992) Somatic sensory responses in the rostral sector of the posterior group (POm) and in the ventral posterior medial nucleus (VPM) of the rat thalamus. J Comp Neurol 318:462–476.

Ding Y, Casagrande VA (1997) The distribution and morphology of LGN K pathway axons within the layers and CO blobs of owl monkey V1. Visual Neurosci 14:691–704.

Ding Y, Casagrande VA (1998) Synaptic and neurochemical characterization of parallel pathways to the cytochrome oxidase blobs of primate visual cortex. J Comp Neurol 391:429–443.

Disney AA, Aoki C, Hawken MJ (2007) Gain modulation by nicotine in macaque V1. Neuron 56:701–713.

Dittman JS, Ryan TA (2019) The control of release probability at nerve terminals. Nat Rev Neurosci 20(3):177–186.

Dobrunz LE, Stevens CF (1997) Heterogeneity of release probability, facilitation, and depletion at central synapses. Neuron 18:995–1008.

Dobrunz LE, Stevens CF (1999) Response of hippocampal synapses to natural stimulation patterns. Neuron 22:157–166.

Douglas RJ, Martin KAC (1991) A functional microcircuit for cat visual cortex. J Physiol (Lond) 440:735–769.

Dreher B, Fukada Y, Rodieck RW (1976) Identification, classification and anatomical secregation of cells with X-like and Y-like properties in the lateral geniculate nucleus of old-world primates. J Physiol (Lond) 258:433–452.

Dreher B, Sefton AJ, Ni SY (1985) The morphology, number, distribution and central projections of Class I retinal ganglion cells in albino and hooded rats. Brain Behav Evol 26:10–48.

Ebert J (2005) Reformation of bird-brain terminology takes off. Nature 433:449.

Economo MN, Viswanathan S, Tasic B, Bas E, Winnubst J, Menon V, Graybuck LT, Nguyen TN, Smith KA, Yao Z, Wang L, Gerfen CR, Chandrashekar J, Zeng H, Looger LL, Svoboda K (2018) Distinct descending motor cortex pathways and their roles in movement. Nature 563:79–84.

Enroth-Cugell C, Robson JG (1966) The contrast sensitivity of retinal ganglion cells of the cat. J Physiol (Lond) 187:517–552.

Erisir A, Van Horn SC, Bickford ME, Sherman SM (1997a) Immunocytochemistry and distribution of parabrachial terminals in the lateral geniculate nucleus of the cat: A comparison with corticogeniculate terminals. J Comp Neurol 377:535–549.

Erisir A, Van Horn SC, Sherman SM (1997b) Relative numbers of cortical and brainstem inputs to the lateral geniculate nucleus. Proc Natl Acad Sci USA 94:1517–1520.

Famiglietti EV, Peters A (1972) The synaptic glomerulus and the intrinsic neuron in the dorsal lateral geniculate nucleus of the cat. J Comp Neurol 144:285–334.

Felch DL, Van Hooser SD (2012) Molecular compartmentalization of lateral geniculate nucleus in the gray squirrel (Sciurus carolinensis). Front Neuroanat 6:12.

Feldman SG, Kruger L (1980) An axonal transport study of the ascending projection of medial lemniscal neurons in the rat. J Comp Neurol 192:427–454.

Felleman DJ, Van Essen DC (1991) Distributed hierarchical processing in the primate cerebral cortex. Cereb Cortex 1:1–47.

Ferster D (1987) Origin of orientation-selective EPSPs in simple cells of cat visual cortex. J Neurosci 7:1780–1791.

Ferster D, Chung S, Wheat H (1996) Orientation selectivity of thalamic input to simple cells of cat visual cortex. Nature 380:249–252.

Findlay JM, Gilchrist ID (2003) Active vision: The psychology of looking and seeing. Oxford: Oxford University Press.

Fischer J, Whitney D (2012) Attention gates visual coding in the human pulvinar. Nat Commun 3:1051.

Fisher TG, Alitto HA, Usrey WM. (2017) Retinal and non-retinal contributions to extraclassical surround suppression in the lateral geniculate nucleus. J Neurosci. 37:226–235.

Fitzpatrick D (1996) The functional organization of local circuits in visual cortex: Insights from the study of tree shrew striate cortex. Cereb Cortex 6:329–341.

Fitzpatrick D, Carey RG, Diamond IT (1980) The projection of the superior colliculus upon the lateral geniculate body in *tupaia glis* and *galago senegalensis*. Brain Res 194:494–499.

Fitzpatrick D, Itoh K, Diamond IT (1983) The laminar organization of the lateral geniculate body and the striate cortex in the squirrel monkey *(saimiri sciureus)*. J Neurosci 3:637–702.

Fitzpatrick D, Usrey WM, Schofield BR, Einstein G (1994) The sublaminar organization of corticogeniculate neurons in layer 6 of macaque striate cortex. Visual Neurosci 11:307–315.

Flight MH (2008) On the probability of release. Nat Rev Neurosci 9:736.

Freund TF, Martin KAC, Whitteridge D (1985) Innervation of cat visual areas 17 and 18 by physiologically identified X- and Y-type thalamic afferents. I. Arborization patterns and quantitative distribution of postsynaptic elements. J Comp Neurol 242:263–274.

Friedlander MJ, Lin C-S, Sherman SM (1979) Structure of physiologically identifed X and Y cells in the cat's lateral geniculate nucleus. Science 204:1114–1117.

Friedlander MJ, Lin C-S, Stanford LR, Sherman SM (1981) Morphology of functionally identified neurons in lateral geniculate nucleus of the cat. J Neurophysiol 46:80–129.

Friedman-Hill SR, Robertson LC, Desimone R, Ungerleider LG (2003) Posterior parietal cortex and the filtering of distractors. Proc Natl Acad Sci USA 100(7):4263–4268.

Fries P (2005) A mechanism for cognitive dynamics: Neuronal communication through neuronal coherence. Trends Cogn Sci 9:474–480.

Fries P (2009) Neuronal gamma-band synchronization as a fundamental process in cortical computation. Annu Rev Neurosci 32:209–224.

Fries P (2015) Rhythms for cognition: Communication through coherence. Neuron 88(1):220–235.

Fukada Y (1971) Receptive field organization of cat optic nerve fibers with special reference to conduction velocity. Vision Res 11:209–226.

Funke K, Nelle E, Li B, Wörgötter F (1996) Corticofugal feedback improves the timing of retino-geniculate signal transmission. NeuroReport 7:2130–2134.

Fuxe K, Dahlstrom A, Hoistad M, Marcellino D, Jansson A, Rivera A, Diaz-Cabiale Z, Jacobsen K, Tinner-Staines B, Hagman B, Leo G, Staines W, Guidolin D, Kehr J, Genedani S, Belluardo N, Agnati LF (2007) From the Golgi-Cajal mapping to the transmitter-based characterization of the neuronal networks leading to two modes of brain communication: Wiring and volume transmission. Brain Res Rev 55(1):17–54.

Gaither NS, Stein BE (1979) Reptiles and mammals use similar sensory organizations in the midbrain. Science 205:595–597.

Gallant JL, Shoup RE, Mazer JA (2000) A human extrastriate area functionally homologous to macaque V4. Neuron 27:227–235.

Garcia-Cabezas MA, Rico B, Sanchez-Gonzalez MA, Cavada C (2007) Distribution of the dopamine innervation in the macaque and human thalamus. Neuroimage 34:965–984.

Garcia-Rill E (2009) Reticular activating system. In: Encyclopedia of neuroscience (Squire LR, ed), pp 137–143. Oxford: Academic Press.

Garthwaite J (2016) From synaptically localized to volume transmission by nitric oxide. J Physiol 594(1):9–18.

Gereau RW, Conn PJ (1994) Potentiation of cAMP responses by metabotropic glutamate receptors depresses excitatory synaptic transmission by a kinase-independent mechanism. Neuron 12:1121–1129.

Gil Z, Connors BW, Amitai Y (1999) Efficacy of thalamocortical and intracortical synaptic connections: Quanta, innervation, and reliability. Neuron 23:385–397.

Gilbert CD (1977) Laminar differences in receptive field properties of cells in cat primary visual cortex. J Physiol (Lond) 268:391–421.

Gilbert CD, Kelly JP (1975) The projections of cells in different layers of the cat's visual cortex. J Physiol (Lond) 163:81–106.

Gilbert CD, Wiesel TN (1979) Morphology and intracortical projections of functionally characterised neurones in the cat visual cortex. Nature 280:120–125.

Godwin DW, Vaughan JW, Sherman SM (1996) Metabotropic glutamate receptors switch visual response mode of lateral geniculate nucleus cells from burst to tonic. J Neurophysiol 76:1800–1816.

Goldberg JH, Fee MS (2012) A cortical motor nucleus drives the basal ganglia-recipient thalamus in singing birds. Nat Neurosci 15:620–627.

Govindaiah, Cox CL (2004) Synaptic activation of metabotropic glutamate receptors regulates dendritic outputs of thalamic interneurons. Neuron 41:611–623.

Govindaiah G, Wang T, Gillette MU, Cox CL (2012) Activity-dependent regulation of retinogeniculate signaling by metabotropic glutamate receptors. J Neurosci 32:12820–12831.

Govindaiah G, Wang T, Gillette MU, Crandall SR, Cox CL (2010) Regulation of inhibitory synapses by presynaptic D(4) dopamine receptors in thalamus. J Neurophysiol 104(5):2757–2765.

Graham J (1977) An autoradiographic study of the efferent connections of the superior colliculus in the cat. J Comp Neurol 173:629–654.

Gray CM, Singer W (1989) Stimulus-specific neuronal oscillations in orientation columns of cat visual cortex. Proc Natl Acad Sci USA 86:1698–1702.

Graybiel AM (1975) Anatomical organization of retinotectal afferents in the cat: An autoradiographic study. Brain Res 96:1–23.

Green DM, Swets JA (1966) Signal detection theory and psychophysics. New York: Wiley.

Grieve KL, Sillito AM (1995) Differential properties of cells in the feline primary visual cortex providing the corticofugal feedback to the lateral geniculate nucleus and visual claustrum. J Neurosci 15:4868–4874.

Groh A, Bokor H, Mease RA, Plattner VM, Hangya B, Stroh A, Deschenes M, Acsady L (2014) Convergence of cortical and sensory driver inputs on single thalamocortical cells. Cereb Cortex 24:3167–3179.

Groleau M, Kang JI, Huppe-Gourgues F, Vaucher E (2015) Distribution and effects of the muscarinic receptor subtypes in the primary visual cortex. Front Synaptic Neurosci 19;7:10.

Gu Z, Liu W, Wei J, Yan Z (2012) Regulation of N-methyl-D-aspartic acid (NMDA) receptors by metabotropic glutamate receptor 7. J Biol Chem 287:10265–10275.

Guido W (2018) Development, form, and function of the mouse visual thalamus. J Neurophysiol 120(1):211–225.

Guido W, Lu S-M, Vaughan JW, Godwin DW, Sherman SM (1995) Receiver operating characteristic (ROC) analysis of neurons in the cat's lateral geniculate nucleus during tonic and burst response mode. Visual Neurosci 12:723–741.

Guido W, Weyand T (1995) Burst responses in thalamic relay cells of the awake behaving cat. J Neurophysiol 74:1782–1786.

Guillery RW (1969a) A quantitative study of synaptic interconnections in the dorsal lateral geniculate nucleus of the cat. Z Zellforsch 96:39–48.

Guillery RW (1969b) The organization of synaptic interconnections in the laminae of the dorsal lateral geniculate nucleus of the cat. Z Zellforsch 96:1–38.

Guillery RW (1970) The laminar distribution of retinal fibers in the dorsal lateral geniculate nucleus of the cat: A new interpretation. J Comp Neurol 138:339–368.

Guillery RW (1995) Anatomical evidence concerning the role of the thalamus in corticocortical communication: A brief review. J Anat 187:583–592.

Guillery RW, Feig SL, Van Lieshout DP (2001) Connections of higher order visual relays in the thalamus: A study of corticothalamic pathways in cats. J Comp Neurol 438:66–85.

Guillery RW, Geisert EE, Polley EH, Mason CA (1980) An analysis of the retinal afferents to the cat's medial interlaminar nucleus and to its rostral thalamic extension, the "geniculate wing." J Comp Neurol 194:117–142.

Guillery RW, Harting JK (2003) Structure and connections of the thalamic reticular nucleus: Advancing views over half a century. J Comp Neurol 463:360–371.

Gulcebi MI, Ketenci S, Linke R, Hacioglu H, Yanali H, Veliskova J, Moshe SL, Onat F, Cavdar S (2011) Topographical connections of the substantia nigra pars reticulata to higher-order thalamic nuclei in the rat. Brain Res Bull 87:312–318.

Gulyás B, Lagae L, Eysel UT, Orban GA (1990) Corticofugal feedback influences the responses of geniculate neurons to moving stimuli. Exp Brain Res 79:441–446.

Guo K, Yamawaki N, Svoboda K, Shepherd GMG (2018) Anterolateral motor cortex connects with a medial subdivision of ventromedial thalamus through cell type-specific circuits, forming an excitatory thalamo-cortico-thalamic Loop via layer 1 apical tuft dendrites of layer 5B pyramidal tract type neurons. J Neurosci 38:8787–8797.

Haas JS, Landisman CE (2011) State-dependent modulation of gap junction signaling by the persistent sodium current. Front Cell Neurosci 5:31.

Halassa MM, Acsady L (2016) Thalamic inhibition: Diverse sources, diverse scales. Trends Neurosci 39:680–693.

Halassa MM, Kastner S (2017) Thalamic functions in distributed cognitive control. Nat Neurosci 20(12):1669–1679.

Halassa MM, Sherman SM (2019) Thalamocortical circuit motifs: A general framework. Neuron 103(5):762–770.

Hamos JE, Van Horn SC, Raczkowski D, Sherman SM (1987) Synaptic circuits involving an individual retinogeniculate axon in the cat. J Comp Neurol 259:165–192.

Hamos JE, Van Horn SC, Raczkowski D, Uhlrich DJ, Sherman SM (1985) Synaptic connectivity of a local circuit neurone in lateral geniculate nucleus of the cat. Nature 317:618–621.

Harding-Forrester S, Feldman DE (2018) Somatosensory maps. Handb Clin Neurol 151:73–102.

Hardingham N, Fox K (2006) The role of nitric oxide and GluR1 in presynaptic and postsynaptic components of neocortical potentiation. J Neurosci 26:7395–7404.

Harting JK, Guillery RW (1976) Organization of retinocollicular pathways in the cat. J Comp Neurol 166(2):133–144.

Harting JK, Hall WC, Diamond IT (1972) Evolution of the pulvinar. Brain Behav Evol 6:424–452.

Harting JK, Updyke BV, Van Lieshout DP (1992) Corticotectal projections in the cat: Anterograde transport studies of twenty-five cortical areas. J Comp Neurol 324:379–414.

Hasse JM, Bragg EM, Murphy AJ, Briggs F (2019) Morphological heterogeneity among corticogeniculate neurons in ferrets: Quantification and comparison with a previous report in macaque monkeys. J Comp Neurol 527(3):546–557.

Hasse JM, Briggs F (2017a) A cross-species comparison of corticogeniculate structure and function. Vis Neurosci 34:E016. doi:10.1017/S095252381700013X. Epub 2017 Nov 16. PMID: 30034107; PMCID: PMC6054447.

Hasse JM, Briggs F (2017b) Corticogeniculate feedback sharpens the temporal precision and spatial resolution of visual signals in the ferret. Proc Natl Acad Sci USA 114(30):E6222–E6230.

Haug H (1987) Brain sizes, surfaces, and neuronal sizes of the cortex cerebri: A stereological investigation of man and his variability and a comparison with some mammals (primates, whales, marsupials, insectivores, and one elephant). Am J Anat 180(2):126–142.

Haverkamp S, Wässle H, Duebel J, Kuner T, Augustine GJ, Feng G, Euler T (2005) The primordial, blue-cone color system of the mouse retina. J Neurosci 25(22):5438–5445.

Headon MP, Sloper JJ, Hiorns RW, Powell TPS (1985) Sizes of neurons in the primate lateral geniculate nucleus during normal development. Dev Brain Res 18:51–56.

Heine WF, Passaglia CL (2011) Spatial receptive field properties of rat retinal ganglion cells. Vis Neurosci 28:403–417.

Helmholtz H (1866) Handbuch der Physiologischen Optik Volume 3. Leipzig: Voss.

Helms MC, Ozen G, Hall WC (2004) Organization of the intermediate gray layer of the superior colliculus. I. Intrinsic vertical connections. J Neurophysiol 91:1706–1715.

Hembrook-Short JR, Mock VL, Usrey WM, Briggs F (2019) Attention enhances the efficacy of communication in V1 local circuits. J Neurosci 39:1066–1076.

Hendrickson AE, Wilson JR, Ogren MP (1978) The neuroanatomical organization of pathways between the dorsal lateral geniculate nucleus and visual cortex in old world and new world promates. J Comp Neurol 182:123–136.

Hendry SHC, Hockfield S, Jones EG, McKay R (1984) Monoclonal antibody that identifies subsets of neurones in the central visual system of monkey and cat. Nature 307:267–269.

Hendry SHC, Reid RC (2000b) The koniocellular pathway in primate vision. Annu Rev Neurosci 23:127–153.

Heukamp AS, Warwick RA, Rivlin-Etzion M (2020) Topographic variations in retinal encoding of visual space. Annu Rev Vis Sci 6:237–259.

Hickey TL, Guillery RW (1974) An autoradiographic study of retinogeniculate pathways in the cat and the fox. J Comp Neurol 156:239–254.

Hickey TL, Guillery RW (1979) Variability of laminar patterns in the human lateral geniculate nucleus. J Comp Neurol 183:221–246.

Hirsch JA, Alonso JM, Reid RC (1995) Visually evoked calcium action potentials in cat striate cortex. Nature 378:612–616.

Hirsch JA, Wang X, Sommer FT, Martinez LM (2015) How inhibitory circuits in the thalamus serve vision. Annu Rev Neurosci 38:309–329.

Hochstein S, Shapley RM (1976) Linear and non-linear subunits in Y cat retinal ganglion cells. J Physiol (Lond) 262:265–284.

Hockfield S, Sur M (1990) Monoclonal antibody cat-301 identifies Y-cells in the dorsal lateral geniculate nucleus of the cat. J Comp Neurol 300:320–330.

Hodgkin AL, Huxley AF (1952) Currents carried by sodium and potassium ions through the membrane of the giant axon of *Loligo*. J Physiol (Lond) 116:449–472.

Hoerder-Suabedissen A, Hayashi S, Upton L, Nolan Z, Casas-Torremocha D, Grant E, Viswanathan S, Kanold PO, Clasca F, Kim Y, Molnar Z (2018) Subset of Cortical Layer 6b Neurons Selectively Innervates Higher Order Thalamic Nuclei in Mice. Cereb Cortex 28(5):1882–1897.

Hoffmann K-P, Stone J, Sherman SM (1972) Relay of receptive-field properties in dorsal lateral geniculate nucleus of the cat. J Neurophysiol 35:518–531.

Hu B (2003) Functional organization of lemniscal and nonlemniscal auditory thalamus. Exp Brain Res 153:543–549.

Hubel DH, Wiesel TN (1959) Receptive fields of single neurones in the cat's striate cortex. J Physiol 148:574–591.

Hubel DH, Wiesel TN (1961) Integrative action in the cat's lateral geniculate body. J Physiol (Lond) 155:385–398.

Hubel DH, Wiesel TN (1962) Receptive fields, binocular interaction and functional architecture in the cat's visual cortex. J Physiol 160:106–154.

Hubel DH, Wiesel TN (1965) Receptive fields and functional architecture in two nonstriate visual areas (18 and 19) of the cat. J Neurophysiol 28:229–289.

Hubel DH, Wiesel TN (1977) Functional architecture of macaque monkey visual cortex. Proc Roy Soc Lond B 198:1–59.

Hubel DH, Wiesel TN (1998) Early exploration of the visual cortex. Neuron 20:401–412.

Hubel DH, Wiesel TN (2005) Brain and visual perception. New York: Oxford University Press.

Hübener M, Shoham D, Grinvald A, Bonhoeffer T (1997) Spatial relationships among three columnar systems in cat area 17. J Neurosci 17:9270–9284.

Huberman AD, Wei W, Elstrott J, Stafford BK, Feller MB, Barres BA (2009) Genetic identification of an On-Off direction-selective retinal ganglion cell subtype reveals a layer-specific subcortical map of posterior motion. Neuron 62(3):327–334.

Hughes A (1979) A rose by any other name . . . on "naming of neurones" by Rowe and Stone. Brain Behav Evol 16:52–64.

Huguenard JR, McCormick DA (1992) Simulation of the currents involved in rhythmic oscillations in thalamic relay neurons. J Neurophysiol 68:1373–1383.

Huguenard JR, McCormick DA (1994) Electrophysiology of the neuron. New York: Oxford University Press.

Huguenard JR, Prince DA (1992) A novel T-type current underlies prolonged Ca2+-dependent burst firing in GABAergic neurons of rat thalamic reticular nucleus. J Neurosci 12:3804–3817.

Humphrey AL, Sur M, Uhlrich DJ, Sherman SM (1985a) Projection patterns of individual X- and Y-cell axons from the lateral geniculate nucleus to cortical area 17 in the cat. J Comp Neurol 233:159–189.

Humphrey AL, Sur M, Uhlrich DJ, Sherman SM (1985b) Termination patterns of individual X- and Y-cell axons in the visual cortex of the cat: Projections to area 18, to the 17-18 border region, and to both areas 17 and 18. J Comp Neurol 233:190–212.

Ichida JM, Casagrande VA (2002) Organization of the feedback pathway from striate cortex (V1) to the lateral geniculate nucleus (LGN) in the owl monkey (*Aotus trivirgatus*). J Comp Neurol 454:272–283.

Ichida JM, Mavity-Hudson JA, Casagrande VA (2014) Distinct patterns of corticogeniculate feedback to different layers of the lateral geniculate nucleus. Eye Brain 2014:57–73.

Ichida JM, Rosa MGP, Casagrande VA (2000) Does the visual system of the flying fox resemble that of primates? the distribution of calcium-binding proteins in the primary visual pathway of *Pteropus poliocephalus*. J Comp Neurol 417:73–87.

Ikeda H, Wright MJ (1972) The periphery effect and its relation to the receptive filed organization of "transient" retinal ganglion cells. J Physiol (Lond) 226:81–82P.

Irvin GE, Casagrande VA, Norton TT (1993) Center/surround relationships of magnocellular, parvocellular, and koniocellular relay cells in primate lateral geniculate nucleus. Visual Neurosci 10:363–373.

Irvin GE, Norton TT, Sesma MA, Casagrande VA (1986) W-like response properties of interlaminar zone cells in the lateral geniculate nucleus of a primate (*Galago crassicaudatus*). Brain Res 362:254–270.

Iwai L, Ohashi Y, van der List D, Usrey WM, Miyashita Y, Kawasaki H (2013) FoxP2 is a parvocellular-specific transcription factor in the visual thalamus of monkeys and ferrets. Cereb Cortex 23(9):2204–2212.

Jack JJB, Noble D, Tsien RW (1975) Electric current flow in excitable cells. Oxford University Press.

Jack JJB, Noble D, Tsien RW (1983) Electric current flow in excitable cells. Oxford: Oxford University Press.

Jacobs GH, Deegan JF, II, Neitz J, Crognale MA, Neitz M (1993) Photopigments and color vision in the nocturnal monkey, *Aotus*. Vision Res 33:1773–1783.

Jacobs GH, Neitz M, Neitz J (1996) Mutations in S-cone pigment genes and the absence of colour vision in two species of nocturnal primate. Proc Biol Sci 263(1371):705–710.

Jahnsen H, Llinás R (1984a) Electrophysiological properties of guinea-pig thalamic neurones: An *in vitro* study. J Physiol (Lond) 349:205–226.

Jahnsen H, Llinás R (1984b) Ionic basis for the electroresponsiveness and oscillatory properties of guinea-pig thalamic neurones *in vitro*. J Physiol (Lond) 349:227–247.

Janssen J, Aleman-Gomez Y, Reig S, Schnack HG, Parellada M, Graell M, Moreno C, Moreno D, Mateos-Perez JM, Udias JM, Arango C, Desco M (2012) Regional specificity of thalamic volume deficits in male adolescents with early-onset psychosis. Br J Psychiatry 200:30–36.

Jiang X, Shen S, Cadwell CR, Berens P, Sinz F, Ecker AS, Patel S, Tolias AS (2015) Principles of connectivity among morphologically defined cell types in adult neocortex. Science 350(6264):aac9462.

Johnston D, Magee JC, Colbert CM, Christie BR (1996) Active properties of neuronal dendrites. Annu Rev Neurosci 19:165–186.

Jones EG (1998) Viewpoint: The core and matrix of thalamic organization. Neurosci 85:331–345.

Jones EG (2001) The thalamic matrix and thalamocortical synchrony. Trends Neurosci 24:595–601.

Jones EG (2007) The thalamus: Second edition. Cambridge: Cambridge University Press.

Jones EG, Wise SP, Coulter JD (1979) Differential thalamic relationships of sensory-motor and parietal cortical fields in monkeys. J Comp Neurol 183:833–882.

Jones HE, Andolina IM, Ahmed B, Shipp SD, Clements JT, Grieve KL, Cudeiro J, Salt TE, Sillito AM (2012) Differential feedback modulation of center and surround mechanisms in parvocellular cells in the visual thalamus. J Neurosci 32(45):15946–15951.

Jones HE, Andolina IM, Oakely NM, Murphy PC, Sillito AM (2000) Spatial summation in lateral geniculate nucleus and visual cortex. Exp Brain Res 135:279–284.

Kaas JH, Baldwin MKL (2019) The evolution of the pulvinar complex in primates and its role in the dorsal and ventral streams of cortical processing. Vision (Basel) 4:1–19.

Kaas JH, Huerta MF, Weber JT, Harting JK (1978) Patterns of retinal terminations and laminar organization of the lateral geniculate nucleus of primates. J Comp Neurol 182:517–554.

Kaas JH, Lyon DC (2007) Pulvinar contributions to the dorsal and ventral streams of visual processing in primates. Brain Res Rev 55:285–296.

Kakei S, Na J, Shinoda Y (2001) Thalamic terminal morphology and distribution of single corticothalamic axons originating from layers 5 and 6 of the cat motor cortex. J Comp Neurol 437:170–185.

Kandel ER, Schwartz JH, Jessell TM (2000) Principles of neural science. New York: McGraw Hill.

Kandler K, Katz LC (1998) Relationship between dye coupling and spontaneous activity in developing ferret visual cortex. Dev Neurosci 20:59–64.

Kaplan E, Purpura K, Shapley RM (1987) Contrast affects the transmission of visual information through the mammalian lateral geniculate nucleus. J Physiol (Lond) 391:267–288.

Kaplan E, Shapley RM (1982) X and Y cells in the lateral geniculate nucleus of macaque monkeys. J Physiol (Lond) 330:125–143.

Kaplan E, Shapley RM (1984) The origin of the S (slow) potential in the mammalian lateral geniculate nucleus. Exp Brain Res 55:111–116.

Kara P, Reinagel P, Reid RC (2000) Low response variability in simultaneously recorded retinal, thalamic, and cortical neurons. Neuron 27:635–646.

Katz LC (1987) Local circuitry of identified projection neurons in cat visual cortex brain slices. J Neurosci 7:1223–1249.

Kavalali ET (2015) The mechanisms and functions of spontaneous neurotransmitter release. Nat Rev Neurosci 16(1):5–16.

Kawamura S, Fukushima N, Hattori S (1979) Topographical origin and ganglion cell type of the retinopulvinar projection in the cat. Brain Res 173(3):419–429.

Kelly JP, Gilbert CD (1975) The projections of different morphological types of ganglion cells in cat retina. J Comp Neurol 163:65–80.

Kelly LR, Li J, Carden WB, Bickford ME (2003) Ultrastructure and synaptic targets of tectothalamic terminals in the cat lateral posterior nucleus. J Comp Neurol 464:472–486.

Kennedy MB (2000) Signal-processing machines at the postsynaptic density. Science 290:750–754.

Kerschensteiner D, Guido W (2017) Organization of the dorsal lateral geniculate nucleus in the mouse. Vis Neurosci 34:E008.

Kim CH, Lee J, Lee JY, Roche KW (2008) Metabotropic glutamate receptors: Phosphorylation and receptor signaling. J Neurosci Res 86:1–10.

Kim J, Kim Y, Nakajima R, Shin A, Jeong M, Park AH, Jeong Y, Jo S, Yang S, Park H, Cho SH, Cho KH, Shim I, Chung JH, Paik SB, Augustine GJ, Kim D (2017) Inhibitory basal ganglia inputs induce excitatory motor signals in the thalamus. Neuron 95(5):1181–1196.

Kim J, Matney CJ, Blankenship A, Hestrin S, Brown SP (2014) Layer 6 corticothalamic neurons activate a cortical output layer, layer 5a. J Neurosci 34:9656–9664.

Kim J, Matney CJ, Roth RH, Brown SP (2016) Synaptic organization of the neuronal circuits of the claustrum. J Neurosci 36:773–784.

Kirchgessner MA, Franklin AD, Callaway EM (2020) Context-dependent and dynamic functional influence of corticothalamic pathways to first- and higher-order visual thalamus. Proc Natl Acad Sci USA 117(23):13066–13077.

Kita T, Kita H (2012) The subthalamic nucleus is one of multiple innervation sites for long-range corticofugal axons: A single-axon tracing study in the rat. J Neurosci 32:5990–5999.

Kölliker A (1896) Handbuch der Gerwebelehre des Menschen. Nervensystemen des Menschen und der Thiere. Leipzig: Engelmann.

Koshland DE, Jr. (1992) The molecule of the year. Science 258(5090):1861.

Krahe TE, El-Danaf RN, Dilger EK, Henderson SC, Guido W (2011) Morphologically distinct classes of relay cells exhibit regional preferences in the dorsal lateral geniculate nucleus of the mouse. J Neurosci 31:17437–17448.

Kraus N, Nicol T (2005) Brainstem origins for cortical "what" and "where" pathways in the auditory system. Trends Neurosci 28:176–181.

Krauzlis RJ, Bogadhi AR, Herman JP, Bollimunta A (2018) Selective attention without a neocortex. Cortex 102:161–175.

Krauzlis RJ, Lovejoy LP, Zenon A (2013) Superior colliculus and visual spatial attention. Annu Rev Neurosci 36:165–182.

Krubitzer L (2007) The magnificent compromise: Cortical field evolution in mammals. Neuron 56(2):201–208.

Krubitzer LA, Kaas JH (1992) The somatosensory thalamus of monkeys: Cortical connections and a redefinition of nuclei in marmosets. J Comp Neurol 319:123–140.

Kuffler S (1952) Neurons in the retina; organization, inhibition and excitation problems. Cold Spring Harb Symp Quant Biol 17:281–292.

Kuffler S (1953) Discharge patterns and functional organization of mammalian retina. J Neurophysiol 16:37–68.

Kullmann DM (2001) Presynaptic kainate receptors in the hippocampus: Slowly emerging from obscurity. Neuron 32:561–564.

Kultas-Ilinsky K, Sivan-Loukianova E, Ilinsky IA (2003) Reevaluation of the primary motor cortex connections with the thalamus in primates. J Comp Neurol 457:133–158.

Kunzle H (1995) Regional and laminar distribution of cortical neurons projecting to either superior or inferior colliculus in the hedgehog tenrec. Cereb Cortex 5(4):338–352.

Kuramoto E, Fujiyama F, Nakamura KC, Tanaka Y, Hioki H, Kaneko T (2011) Complementary distribution of glutamatergic cerebellar and GABAergic basal ganglia afferents to the rat motor thalamic nuclei. Eur J Neurosci 33:95–109.

Kuramoto E, Ohno S, Furuta T, Unzai T, Tanaka YR, Hioki H, Kaneko T (2013) Ventral medial nucleus neurons send thalamocortical afferents more widely and more preferentially to layer 1 than neurons of the ventral anterior-ventral lateral nuclear complex in the rat. Cereb Cortex 25:221–235.

Kuramoto E, Pan S, Furuta T, Tanaka YR, Iwai H, Yamanaka A, Ohno S, Kaneko T, Goto T, Hioki H (2017) Individual mediodorsal thalamic neurons project to multiple areas of the rat prefrontal cortex: A single neuron-tracing study using virus vectors. J Comp Neurol 525:166–185.

Lachica EA, Casagrande VA (1992) Direct W-like geniculate projections to the cytochrome oxidase (CO) blobs in primate visual cortex: Axon morphology. J Comp Neurol 319:141–158.

Lam YW, Cox CL, Varela C, Sherman SM (2005) Morphological correlates of triadic circuitry in the lateral geniculate nucleus of cats and rats. J Neurophysiol 93:748–757.

Lam YW, Nelson CS, Sherman SM (2006) Mapping of the functional interconnections between reticular neurons using photostimulation. J Neurophysiol 96:2593–2600.

Lam YW, Sherman SM (2010) Functional organization of the somatosensory cortical layer 6 feedback to the thalamus. Cereb Cortex 20:13–24.

Lam YW, Sherman SM (2013) Activation of both Group I and Group II metabotropic glutamatergic receptors suppress retinogeniculate transmission. Neurosci 242:78–84.

Lam YW, Sherman SM (2019) Convergent synaptic inputs to layer 1 cells of mouse cortex. Eur J Neurosci 49:1399.

Landisman CE, Connors BW (2005) Long-term modulation of electrical synapses in the mammalian thalamus. Science 310:1809–1813.

Landò L, Zucker RS (1994) Ca2+ cooperativity in neurosecretion measured using photolabile Ca2+ chelators. J Neurophysiol 72:825–830.

Larkman AU, Major G, Stratford KJ, Jack JJ (1992) Dendritic morphology of pyramidal neurones of the visual cortex of the rat. IV: Electrical geometry. J Comp Neurol 323(2):137–152.

Larkum ME, Senn W, Luscher HR (2004) Top-down dendritic input increases the gain of layer 5 pyramidal neurons. Cereb Cortex 14:1059–1070.

Larkum ME, Waters J, Sakmann B, Helmchen F (2007) Dendritic spikes in apical dendrites of neocortical layer 2/3 pyramidal neurons. J Neurosci 27:8999–9008.

Larkum ME, Zhu JJ, Sakmann B (1999) A new cellular mechanism for coupling inputs arriving at different cortical layers. Nature 398:338–341.

Lavallée P, Urbain N, Dufresne C, Bokor H, Acsády L, Deschênes M (2005) Feedforward inhibitory control of sensory information in higher-order thalamic nuclei. J Neurosci 25:7489–7498.

Le Gros Clark WE (1941a) The laminar organization and cell content of the lateral geniculate body in the monkey. J Anat 75(Pt 4):419–433.

Le Gros Clark WE (1941b) The lateral geniculate body in the platyrrhine monkeys. J Anat 76(Pt 1):131–140.

Lee BB (1996) Receptive field structure in the primate retina. Vision Res 36:631–644.

Lee CC, Sherman SM (2008) Synaptic properties of thalamic and intracortical inputs to layer 4 of the first- and higher-order cortical areas in the auditory and somatosensory systems. J Neurophysiol 100:317–326.

Lee CC, Sherman SM (2009a) Glutamatergic inhibition in sensory neocortex. Cereb Cortex 19:2281–2289.

Lee CC, Sherman SM (2009b) Modulator property of the intrinsic cortical projection from layer 6 to layer 4. Front Syst Neurosci 3:1–5.

Lee CC, Sherman SM (2010) Topography and physiology of ascending streams in the auditory tectothalamic pathway. Proc Nat Acad Sci USA 107:372–377.

Lee J, Maunsell JHR (2010) Attentional modulation of MT neurons with single or multiple stimuli in their receptive fields. J Neurosci 30:3058–3066.

Lee SH, Dan Y (2012) Neuromodulation of brain states. Neuron 76:209–222.

Lehmkuhle S, Sherman SM, Kratz KE (1984) Spatial contrast sensitivity of dark-reared cats with striate cortex lesions. J Neurosci 4:2419–2424.

Lennie P (1980) Perceptual signs of parallel pathways. Philos Trans R Soc Lond B Biol Sci 290:23–37.

Lesica NA, Stanley GB (2004) Encoding of natural scene movies by tonic and burst spikes in the lateral geniculate nucleus. J Neurosci 24:10731–10740.

LeVay S (1986) Synaptic organization of claustral and geniculate afferents to the visual cortex of the cat. J Neurosci 6(12):3564–3575.

LeVay S, Gilbert CD (1976) Laminar patterns of geniculocortical projection in the cat. Brain Res 113:1–19.

LeVay S, McConnell SK (1982) ON and OFF layers in the lateral geniculate nucleus of the mink. Nature 300:350–351.

LeVay S, Sherk H (1981) The visual claustrum of the cat. I. Structure and connections. J Neurosci 1:956–980.

Levenson DH, Fernandez-Duque E, Evans S, Jacobs GH (2007) Mutational changes in S-cone opsin genes common to both nocturnal and cathemeral Aotus monkeys. Am J Primatol 69:757–765.

Leventhal AG, Keens J, Törk I (1980) The afferent ganglion cells and cortical projections of the retinal recipient zone (RRZ) of the cat's "pulvinar complex." J Comp Neurol 194:535–554.

Leventhal AG, Rodieck RW, Dreher B (1981) Retinal ganglion cell classes in the old world monkey: Morphology and central projections. Science 213:1139–1142.

Leventhal AG, Rodieck RW, Dreher B (1985) Central projections of cat retinal ganglion cells. J Comp Neurol 237:216–226.

Levesque M, Charara A, Gagnon S, Parent A, Deschênes M (1996) Corticostriatal projections from layer V cells in rat are collaterals of long-range corticofugal axons. Brain Res 709:311–315.

Levick WR, Cleland BG, Dubin MW (1972) Lateral geniculate neurons of cat: Retinal inputs and physiology. Invest Ophthalmol 11:302–311.

Levine MW, Cleland BG (2001) An analysis of the effect of retinal ganglion cell impulses upon the firing probability of neurons in the dorsal lateral geniculate nucleus of the cat. Brain Res 902:244–254.

Levitt JB, Schumer RA, Sherman SM, Spear PD, Movshon JA (2001) Visual response properties of neurons in the LGN of normally reared and visually deprived macaque monkeys. J Neurophysiol 85:2111–2129.

Levitt JB, Yoshioka T, Lund JS (1995) Connections between the pulvinar complex and cytochrome oxidase-defined compartments in visual area V2 of macaque monkey. Exp Brain Res 104:419–430.

Levy RB, Reyes AD (2012) Spatial profile of excitatory and inhibitory synaptic connectivity in mouse primary auditory cortex. J Neurosci 32:5609–5619.

Li J, Guido W, Bickford ME (2003a) Two distinct types of corticothalamic EPSPs and their contribution to short-term synaptic plasticity. J Neurophysiol 90:3429–3440.

Li JL, Wang ST, Bickford ME (2003b) Comparison of the ultrastructure of cortical and retinal terminals in the rat dorsal lateral geniculate and lateral posterior nuclei. J Comp Neurol 460:394–409.

Li X, Yu B, Sun Q, Zhang Y, Ren M, Zhang X, Li A, Yuan J, Madisen L, Luo Q, Zeng H, Gong H, Qiu Z (2018) Generation of a whole-brain atlas for the cholinergic system and mesoscopic projectome analysis of basal forebrain cholinergic neurons. Proc Natl Acad Sci USA 115(2):415–420.

Li Y, et al. (2020) Distinct subnetworks of the thalamic reticular nucleus. Nature 583(7818):819–824.

Linden DC, Guillery RW, Cucchiaro J (1981) The dorsal lateral geniculate nucleus of the normal ferret and its postnatal development. J Comp Neurol 203(2):189–211.

Linden R, Perry VH (1983) Massive retinotectal projection in rats. Brain Res 272:145–149.

Lisman JE (1997) Bursts as a unit of neural information: Making unreliable synapses reliable. Trends Neurosci 20:38–43.

Litvina EY, Chen C (2017) Functional convergence at the retinogeniculate synapse. Neuron 96:330–338.

Liu XB, Jones EG (2003) Fine structural localization of connexin-36 immunoreactivity in mouse cerebral cortex and thalamus. J Comp Neurol 466:457–467.

Llano DA, Sherman SM (2008) Evidence for nonreciprocal organization of the mouse auditory thalamocortical-corticothalamic projection systems. J Comp Neurol 507:1209–1227.

Llano DA, Sherman SM (2009) Differences in intrinsic properties and local network connectivity of identified layer 5 and layer 6 adult mouse auditory corticothalamic neurons support a dual corticothalamic projection hypothesis. Cereb Cortex 19:2810–2826.

Llinás R (1988) The intrinsic electrophysiological properties of mammalian neurons: Insights into central nervous system. Science 242:1654–1664.

Llinás RR, Steriade M (2006) Bursting of thalamic neurons and states of vigilance. J Neurophysiol 95:3297–3308.

Lo FS, Ziburkus J, Guido W (2002) Synaptic mechanisms regulating the activation of a Ca^{2+}-mediated plateau potential in developing relay cells of the LGN. J Neurophysiol 87:1175–1185.

Lo F-S, Lu S-M, Sherman SM (1991) Intracellular and extracellular in vivo recording of different response modes for relay cells of the cat's lateral geniculate nucleus. Exp Brain Res 83:317–328.

Lovejoy LP, Krauzlis RJ (2010) Inactivation of primate superior colliculus impairs covert selection of signals for perceptual judgments. Nat Neurosci 13:261–U153.

Lu GW, Willis WD, Jr. (1999) Branching and/or collateral projections of spinal dorsal horn neurons. Brain Res Rev 29:50–82.

Lujan R, Nusser Z, Roberts JD, Shigemoto R, Somogyi P (1996) Perisynaptic location of metabotropic glutamate receptors mGluR1 and mGluR5 on dendrites and dendritic spines in the rat hippocampus. Eur J Neurosci 8:1488–1500.

Lund JS (1973) Organization of neurons in the visual cortex, area 17, of the monkey (*Macaca mulatta*). J Comp Neurol 147:455–496.

Lund JS (1988) Anatomical organization of macaque monkey striate visual cortex. Annu Rev Neurosci 11:253–288.

Lund JS, Boothe RG (1975) Interlaminar connections and pyramidal neuron organisation in the visual cortex, area 17, of the Macaque monkey. J Comp Neurol 159:304–334.

Lund JS, Lund RD, Hendrickson AE, Bunt AH, Fuchs AF (1975) The origin of efferent pathways from the primary visual cortex, area 17, of the macaque monkey as shown by retrograde transport of horseradish peroxidase. J Comp Neurol 164:287–303.

Luscher C, Huber KM (2010) Group 1 mGluR-dependent synaptic long-term depression: Mechanisms and implications for circuitry and disease. Neuron 65:445–459.

Ma Y, Hioki H, Konno M, Pan S, Nakamura H, Nakamura KC, Furuta T, Li JL, Kaneko T (2011) Expression of gap junction protein connexin36 in multiple subtypes of GABAergic neurons in adult rat somatosensory cortex. Cereb Cortex 21:2639–2649.

MacDermott AB, Role LW, Siegelbaum SA (1999) Presynaptic ionotropic receptors and the control of transmitter release. Annu Rev Neurosci 22:443–485.

MacLeod NK, James TA, Kilpatrick IC, Starr MS (1980) Evidence for a GABAergic nigrothalamic pathway in the rat. II. Electrophysiological studies. Exp Brain Res 40(1):55–61.

Macmillan NA, Creelman CD (1991) Detection theory: A user's guide. Cambridge: Cambridge University Press.

Magee JC, Cook EP (2000) Somatic EPSP amplitude is independent of synapse location in hippocampal pyramidal neurons. Nat Neurosci 3:895–903.

Magoun HW (1952) An ascending reticular activating system in the brain stem. Arch Neurol Psychiatry 67:145–154.

Marshel JH, Kaye AP, Nauhaus I, Callaway EM (2012) Anterior-posterior direction opponency in the superficial mouse lateral geniculate nucleus. Neuron 76:713–720.

Martin PR, Lee BB, White AJR, Solomon SG, Rüttiger L (2001) Chromatic sensitivity of ganglion cells in the peripheral primate retina. Nature 410:933–936.

Martin PR, White AJ, Goodchild AK, Wilder HD, Sefton AE (1997) Evidence that blue-on cells are part of the third geniculocortical pathway in primates. Eur J Neurosci 9:1536–1541.

Martinez-Garcia RI, Voelcker B, Zaltsman JB, Patrick SL, Stevens TR, Connors BW, Cruikshank SJ (2020) Two dynamically distinct circuits drive inhibition in the sensory thalamus. Nature 10–2512.

Masland RH (2011) Cell populations of the retina: The Proctor lecture. Invest Ophthalmol Vis Sci 52(7):4581–4591.

Masland RH (2012) The neuronal organization of the retina. Neuron 76(2):266–280.

Masterson SP, Li J, Bickford ME (2009) Synaptic organization of the tectorecipient zone of the rat lateral posterior nucleus. J Comp Neurol 515:647–663.

Mastronarde DN (1987) Two classes of single-input X-cells in cat lateral geniculate nucleus. II. Retinal inputs and the generation of receptive-field properties. J Neurophysiol 57:381–413.

Maunsell JHR, Treue S (2006) Feature-based attention in visual cortex. Trends Neurosci 29(6):317–322.

Maunsell JHR (2015) Neuronal mechanisms of visual attention. Annu Rev Vis Sci 1:373–391.

Maunsell JHR, Ghose GM, Assad JA, McAdams CJ, Boudreau CE, Noerager BD (1999) Visual response latencies of magnocellular and parvocellular LGN neurons in macaque monkeys. Visual Neurosci 16:1–14.

McAdams CJ, Reid RC (2005) Attention modulates the responses of simple cells in monkey primary visual cortex. J Neurosci 25:11023–11033.

McAlonan K, Brown VJ, Bowman EM (2000) Thalamic reticular nucleus activation reflects attentional gating during classical conditioning. J Neurosci 20:8897–8901.

McAlonan K, Cavanaugh J, Wurtz RH (2006) Attentional modulation of thalamic reticular neurons. J Neurosci 26:4444–4450.

McAlonan K, Cavanaugh J, Wurtz RH (2008) Guarding the gateway to cortex with attention in visual thalamus. Nature 456:391–394.

McClurkin JW, Optican LM, Richmond BJ (1994) Cortical feedback increases visual information transmitted by monkey parvocellular lateral geniculate nucleus neurons. Visual Neurosci 11:601–617.

McCormick DA (1989) Cholinergic and noradrenergic modulation of thalamocortical processing. Trends Neurosci 12:215–221.

McCormick DA (1992) Neurotransmitter actions in the thalamus and cerebral cortex and their role in neuromodulation of thalamocortical activity. Prog Neurobiol 39:337–388.

McCormick DA (2014) Membrane potential and action potential. In: Fundamental neuroscience (Squire L, Berg D, Bloom FE, du Lac S, Ghosh A, Spitzer N, eds), pp 94–116. Oxford: Academic Press.

McCormick DA, Huguenard JR (1992) A model of the electrophysiological properties of thalamocortical relay neurons. J Neurophysiol 68:1384–1400.

McCormick DA, Pape H-C (1988) Acetycholine inhibits identified interneurons in the cat lateral geniculate nucleus. Nature 334:246–248.

McCormick DA, Prince DA (1986) Acetylcholine induces burst firing in thalamic reticular neurones by activating a potassium conductance. Nature 319:402–405.

McFarland NR, Haber SN (2002) Thalamic relay nuclei of the basal ganglia form both reciprocal and nonreciprocal cortical connections, linking multiple frontal cortical areas. J Neurosci 22:8117–8132.

McGinley MJ, Vinck M, Reimer J, Batista-Brito R, Zagha E, Cadwell CR, Tolias AS, Cardin JA, McCormick DA (2015) Waking state: Rapid variations modulate neural and behavioral responses. Neuron 87:1143–1161.

McHaffie JG, Stanford TR, Stein BE, Coizet W, Redgrave P (2005) Subcortical loops through the basal ganglia. Trends Neurosci 28:401–407.

Means LW, Harrell TH, Mayo ES, Alexander GB (1974) Effects of dorsomedial thalamic lesions on spontaneous alternation, maze, activity and runway performance in the rat. Physiol Behav 12:973–979.

Merabet L, Desautels A, Minville K, Casanova C (1998) Motion integration in a thalamic visual nucleus. Nature 396:265–268.

Merigan WH, Maunsell JHR (1993) How parallel are the primate visual pathways. Annu Rev Neurosci 16:369–402.

Merrick C, Godwin CA, Geisler MW, Morsella E (2014) The olfactory system as the gateway to the neural correlates of consciousness. Front Psychol 4:1011.

Miller AJ, Sherman SM (2019) Mouse S1 and V1 each initiates a feedback circuit to itself via higher-order thalamus (abstract). Program No 221 10 2019 Neuroscience Meeting Planner Chicago, IL: Society for Neuroscience, 2019. https://www.abstractsonline.com/pp8/#!/7883/presentation/62165

Miller EK, Buschman TJ (2013) Cortical circuits for the control of attention. Curr Opin Neurobiol 23(2):216–222.

Miller KD (2016) Canonical computations of cerebral cortex. Curr Opin Neurobiol 37:75–84.

Miller LM, Escabi MA, Read HL, Schreiner CE (2001a) Functional convergence of response properties in the auditory thalamocortical system. Neuron 32(1):151–160.

Miller LM, Escabí MA, Schreiner CE (2001b) Feature selectivity and interneuronal cooperation in the thalamocortical system. J Neurosci 21:8136–8144.

Miller RJ (1998) Presynaptic receptors. Annu Rev Pharmacol Toxicol 38:201–227.

Mineault PJ, Tring E, Trachtenberg JT, Ringach DL (2016) Enhanced spatial resolution during locomotion and heightened attention in mouse primary visual cortex. J Neurosci 36:6382–6392.

Mitchell AS (2015) The mediodorsal thalamus as a higher order thalamic relay nucleus important for learning and decision-making. Neurosci Biobehav Rev 54:76–88.

Mo C, Petrof I, Viaene AN, Sherman SM (2017) Synaptic properties of the lemniscal and paralemniscal pathways to the mouse somatosensory thalamus. Proc Natl Acad Sci USA 114:E6212–E6221.

Mo C, Sherman SM (2019) A sensorimotor pathway via higher-order thalamus. J Neurosci 39:692–704.

Molotchnikoff S, Tremblay F, Lepore F (1984) The role of the visual cortex in response properties of lateral geniculate cells in rats. Exp Brain Res 53:223–232.

Morgan JL, Berger DR, Wetzel AW, Lichtman JW (2016) The fuzzy logic of network connectivity in mouse visual thalamus. Cell 165:192–206.

Moruzzi G, Magoun HW (1949) Brain stem reticular formation and activation of the EEG. Electroencephalogr Clin Neurophysiol 1:455–473.

Mukherjee P, Kaplan E (1995) Dynamics of neurons in the cat lateral geniculate nucleus: In vivo electrophysiology and computational modeling. J Neurophysiol 74:1222–1243.

Mundy NI, Morningstar NC, Baden AL, Fernandez-Duque E, Davalos VM, Bradley BJ (2016) Can colour vision re-evolve? Variation in the X-linked opsin locus of cathemeral Azara's owl monkeys (*Aotus azarae azarae*). Front Zool 13:9.

Murphy AJ, Shaw L, Hasse JM, Goris RLT, Briggs F (2020) Optogenetic activation of corticogeniculate feedback stabilizes response gain and increases information coding in LGN neurons. J Comput Neurosci 10-00754.

Nakamura H, Hioki H, Furuta T, Kaneko T (2015) Different cortical projections from three subdivisions of the rat lateral posterior thalamic nucleus: A single-neuron tracing study with viral vectors. Eur J Neurosci 41:1294–1310.

Nakamura S, Narumi T, Tsutsui K, Iijima T (2009) Difference in the functional significance between the lemniscal and paralemniscal pathways in the perception of direction of single-whisker stimulation examined by muscimol microinjection. Neurosci Res 64:323–329.

Nassi JJ, Callaway EM (2009) Parallel processing strategies of the primate visual system. Nat Rev Neurosci 10:360–372.

Naumann RK (2015) Even the smallest mammalian brain has yet to reveal its secrets. Brain Behav Evol 85(1):1–3.

Netter FH (1977) The Ciba collection of medical illustrations, vol. 1: Nervous system: A compilation of paintings on the normal and pathologic anatomy of the nervous system, with a supplement on the hypothalamus. Summit, NJ: CIBA.

Nicoll RA, Malenka RC, Kauer JA (1990) Functional comparison of neurotransmitter receptor subtypes in mammalian central nervous system. Physiol Rev 70:513–565.

Nobre K, Nobre A, Kastner S (2014) The Oxford handbook of attention. Oxford: Oxford University Press.

Norton TT, Casagrande VA, Irvin GE, Sesma MA, Petry HM (1988) Contrast-sensitivity functions of W-, X-, and Y-like relay cells in the lateral geniculate nucleus of bush baby, *Galago crassicaudatus*. J Neurophysiol 59:1639–1656.

Nusser Z, Mulvihill E, Streit P, Somogyi P (1994) Subsynaptic segregation of metabotropic and ionotropic glutamate receptors as revealed by immunogold localization. Neurosci 61:421–427.

O'Connor DH, Fukui MM, Pinsk MA, Kastner S (2002) Attention modulates responses in the human lateral geniculate nucleus. Nat Neurosci 5(11):1203–1209.

Ohishi H, Shigemoto R, Nakanishi S, Mizuno N (1993) Distribution of the messenger RNA for a metabotropic glutamate receptor, mGluR2, in the central nervous system of the rat. Neurosci 53:1009–1018.

Ojima H, Murakami K, Kishi K (1996) Dual termination modes of corticothalamic fibers originating from pyramids of layers 5 and 6 in cat visual cortical area 17. Neurosci Lett 208:57–60.

Olsen SR, Bortone DS, Adesnik H, Scanziani M (2012) Gain control by layer six in cortical circuits of vision. Nature 483:47–52.

Osborne LC, Hohl SS, Bialek W, Lisberger SG (2007) Time course of precision in smooth-pursuit eye movements of monkeys. J Neurosci 27(11):2987–2998.

Otani S, Auclair N, Desce JM, Roisin MP, Crepel F (1999) Dopamine receptors and groups I and II mGluRs cooperate for long-term depression induction in rat prefrontal cortex through converging postsynaptic activation of MAP kinases. J Neurosci 19:9788–9802.

Overstreet-Wadiche L, McBain CJ (2015) Neurogliaform cells in cortical circuits. Nat Rev Neurosci 16(8):458–468.

Ozen G, Augustine GJ, Hall WC (2000) Contribution of superficial layer neurons to premotor bursts in the superior colliculus. J Neurophysiol 84:460–471.

Padberg J, Cerkevich C, Engle J, Rajan AT, Recanzone G, Kaas J, Krubitzer L (2009) Thalamocortical connections of parietal somatosensory cortical fields in macaque monkeys are highly divergent and convergent. Cereb Cortex 19:2038–2064.

Papasavvas CA, Parrish RR, Trevelyan AJ (2020) Propagating activity in neocortex, mediated by gap junctions and modulated by extracellular potassium. eNeuro 7(2): ENEURO.0387-19.2020.

Pape HC, McCormick DA (1995) Electrophysiological and pharmacological properties of interneurons in the cat dorsal lateral geniculate nucleus. Neurosci 68:1105–1125.

Park A, Li Y, Masri R, Keller A (2017) Presynaptic and extrasynaptic regulation of posterior nucleus of thalamus. J Neurophysiol 118(1):507–519.

Parnaudeau S, O'Neill PK, Bolkan SS, Ward RD, Abbas AI, Roth BL, Balsam PD, Gordon JA, Kellendonk C (2013) Inhibition of mediodorsal thalamus disrupts thalamofrontal connectivity and cognition. Neuron 77:1151–1162.

Percival KA, Koizumi A, Masri RA, Buzas P, Martin PR, Grunert U (2014) Identification of a pathway from the retina to koniocellular layer K1 in the lateral geniculate nucleus of marmoset. J Neurosci 34(11):3821–3825.

Perry VH, Cowey A (1984) Retinal ganglion cells that project to the superior colliculus and pretectum in the macaque monkey. Neurosci 12:1125–1137.

Person AL, Perkel DJ (2005) Unitary IPSPs drive precise thalamic spiking in a circuit required for learning. Neuron 46:129–140.

Peters A, Payne BR (1993) Numerical relationships between geniculocortical afferents and pyramidal cell modules in cat primary visual cortex. Cereb Cortex 3:69–78.

Peters A, Payne BR, Budd J (1994) A numerical analysis of the geniculocortical input to striate cortex in the monkey. Cereb Cortex 4:215–229.

Petersen CC (2007) The functional organization of the barrel cortex. Neuron 56:339–355.

Petersen SE, Posner MI (2012) The attention system of the human brain: 20 years after. Annu Rev Neurosci 35:73–89.

Petersen SE, Robinson DL, Keys W (1985) Pulvinar nuclei of the behaving rhesus monkey: Visual responses and their modulation. J Neurophysiol 54:867–886.

Petrof I, Viaene AN, Sherman SM (2012) Two populations of corticothalamic and interareal corticocortical cells in the subgranular layers of the mouse primary sensory cortices. J Comp Neurol 520:1678–1686.

Petrof I, Viaene AN, Sherman SM (2015) Properties of the primary somatosensory cortex projection to the primary motor cortex in the mouse. J Neurophysiol 113:2652.

Pi HJ, Hangya B, Kvitsiani D, Sanders JI, Huang ZJ, Kepecs A (2013) Cortical interneurons that specialize in disinhibitory control. Nature 503(7477):521–524.

Pickel V, Segal M, eds (2013) The synapse: Structure and function. Academic Press.

Pin JP, Bockaert J (1995) Get receptive to metabotropic glutamate receptors. Curr Opin Neurobiol 5:342–349.

Pin JP, Duvoisin R (1995) The metabotropic glutamate receptors: Structure and functions. Neuropharmacol 34:1–26.

Pinault D (2004) The thalamic reticular nucleus: Structure, function and concept. Brain Res Rev 46:1–31.

Piscopo DM, El-Danaf RN, Huberman AD, Niell CM (2013) Diverse visual features encoded in mouse lateral geniculate nucleus. J Neurosci 33:4642–4656.

Pons TP, Kaas JH (1985) Connections of area 2 of somatosensory cortex with the anterior pulvinar and subdivisions of the ventroposterior complex in macaque monkeys. J Comp Neurol 240:16–36.

Porter JT, Nieves D (2004) Presynaptic GABAB receptors modulate thalamic excitation of inhibitory and excitatory neurons in the mouse barrel cortex. J Neurophysiol 92(5):2762–2770.

Posner MI (2012) Cognitive neuroscience of attention. New York: Guildford.

Power BD, Kolmac CI, Mitrofanis J (1999) Evidence for a large projection from the zona incerta to the dorsal thalamus. J Comp Neurol 404:554–565.

Prasad JA, Carroll BJ, Sherman SM (2020) Layer 5 corticofugal projections from diverse cortical areas: Variations on a pattern of thalamic and extra-thalamic targets. J Neurosci 40:5785–5796.

Priebe NJ, Ferster D (2012) Mechanisms of neuronal computation in mammalian visual cortex. Neuron 75(2):194–208.

Przybyszewski AW, Gaska JP, Foote W, Pollen DA (2000) Striate cortex increases contrast gain of macaque LGN neurons. Vis Neurosci 17(4):485–494.

Purves D, Augustine GJ, Fitzpatrick D, Hall WC, Lamantia A-S, White LE (2012) Neuroscience. Sunderland, MA: Sinauer.

Raastad M, Shepherd GM (2003) Single-axon action potentials in the rat hippocampal cortex. J Physiol 548:745–752.

Rafal RD, Posner MI (1987) Deficits in human visual spatial attention following thalamic lesions. Proc Natl Acad Sci USA 84:7349–7353.

Rall W (1969a) Distributions of potential in cylindrical coordinates and time constants for a membrane cylinder. Biophys J 9:1509–1541.

Rall W (1969b) Time constants and electrotonic length of membrane cylinders and neurons. Biophys J 9:1483–1508.

Ralston HJ (1971) Evidence for presynaptic dendrites and a proposal for their mechanism of action. Nature 230:585–587.

Ramcharan EJ, Gnadt JW, Sherman SM (2005) Higher-order thalamic relays burst more than first-order relays. Proc Natl Acad Sci USA 102:12236–12241.

Rao RP, Ballard DH (1999) Predictive coding in the visual cortex: A functional interpretation of some extra-classical receptive-field effects. [see comments]. Nat Neurosci 2:79–87.

Rathbun DL, Usrey WM (2009) The geniculo-striate pathway. In: Encyclopedia of neuroscience (Binder MD, Hirokawa N, Windhost U, Hirsch MC, eds), pp 1707–1710. Heidelberg: Springer-Verlag.

Rathbun DL, Alitto HJ, Warland DK, Usrey WM. (2016) Stimulus contrast and retinogeniculate signal processing. Frontiers in Neural Circuits, 10:8, doi: 10.3389/fncir.2016.00008.

Rathbun DL, Warland DK, Usrey WM (2010) Spike timing and information transmission at retinogeniculate synapses. J Neurosci 30:13558–13566.

Read HL, Nauen DW, Escabã MA, Miller LM, Schreiner CE, Winer JA (2011) Distinct core thalamocortical pathways to central and dorsal primary auditory cortex. Hear Res 274(1–2):95–104.

Reichova I, Sherman SM (2004) Somatosensory corticothalamic projections: Distinguishing drivers from modulators. J Neurophysiol 92:2185–2197.

Reid RC, Alonso JM (1995) Specificity of monosynaptic connections from thalamus to visual cortex. Nature 378:281–284.

Reid RC, Alonso JM, Usrey WM (2001) Integration of thalamic inputs to cat primary visual cortex. In: The cat primary visual cortex (Payne BR, Peters A, eds), pp 310–342. San Diego: Academic Press.

Reid RC, Shapley RM (1992) Spatial structure of cone inputs to receptive fields in primate lateral geniculate nucleus. Nature 356:716–718.

Reid RC, Usrey WM (2004) Functional connectivity in the pathway from retina to visual cortex. In: The visual neurosciences (Chalupa LM and Werner JS, eds), pp 673–679. Cambridge, MA: MIT Press.

Reinagel P, Godwin DW, Sherman SM, Koch C (1999) Encoding of visual information by LGN bursts. J Neurophysiol 81:2558–2569.

Reiner A, et al. (2004) The avian brain nomenclature forum: Terminology for a new century in comparative neuroanatomy. J Comp Neurol 473:E1–E6.

Reyes A, Lujan R, Rozov A, Burnashev N, Somogyi P, Sakmann B (1998) Target-cell-specific facilitation and depression in neocortical circuits. Nat Neurosci 1:279–285.

Reynolds JH, Chelazzi L (2004) Attentional modulation of visual processing. Annu Rev Neurosci 27:611–647.

Rikhye RV, Gilra A, Halassa MM (2018) Thalamic regulation of switching between cortical representations enables cognitive flexibility. Nat Neurosci 21:1753–1763.

Robson JA, Hall WC (1977) The organization of the pulvinar in the grey squirrel (*Sciurus carolinensis*) II. Synaptic organization and comparisons with the dorsal lateral geniculate nucleus. J Comp Neurol 173:389–416.

Rockland KS (1996) Two types of corticopulvinar terminations: Round (type 2) and elongate (type 1). J Comp Neurol 368:57–87.

Rockland KS (1998) Convergence and branching patterns of round, type 2 corticopulvinar axons. J Comp Neurol 390:515–536.

Rockland KS, Pandya DN (1979) Laminar origins and terminations of cortical connections of the occipital lobe in the rhesus monkey. Brain Res 179:3–20.

Rodieck RW, Watanabe M (1993) Survey of the morphology of macaque retinal ganglion cells that project to the pretectum, superior colliculus, and parvicellular laminae of the lateral geniculate nucleus. J Comp Neurol 338:289–303.

Rowe MH, Fischer Q (2001) Dynamic properties of retino-geniculate synapses in the cat. Visual Neurosci 18:219–231.

Rowe MH, Stone J (1977) Naming of Neurones: Classification and naming of cat retinal ganglion cells. Brain Behav Evol 14:185–216.

Roy S, Jayakumar J, Martin PR, Dreher B, Saalmann YB, Hu DP, Vidyasagar TR (2009) Segregation of short-wavelength-sensitive (S) cone signals in the macaque dorsal lateral geniculate nucleus. Eur J Neurosci 30:1517–1526.

Roy SA, Alloway KD (2001) Coincidence detection or temporal integration? What the neurons in somato-sensory cortex are doing. J Neurosci 21:2462–2473.

Saalmann YB, Pinsk MA, Wang L, Li X, Kastner S (2012) The pulvinar regulates information transmission between cortical areas based on attention demands. Science 337:753–756.

Sakai ST, Inase M, Tanji J (1996) Comparison of cerebellothalamic and pallidothalamic projections in the monkey (Macaca fuscata): A double anterograde labeling study. J Comp Neurol 368:215–228.

Salt TE, Eaton SA (1995) Modulation of sensory neurone excitatory and inhibitory responses in the ventrobasal thalamus by activation of metabotropic excitatory amino acid receptors. Neuropharmacol 34:1043–1051.

Sampathkumar V, Miller-Hansen A, Sherman SM, Kasthuri N (2021) An ultrastructural connectomic analysis of a higher order thalamocortical circuit in the mouse. Eur J Neurosci 53:750–762.

Sampathkumar V, Miller-Hansen A, Sherman SM, & Kasthuri N (2021) Integration of signals from different cortical areas in higher order thalamic neurons. Proc Natl Acad Sci U.S.A 118, (30) available from: PM:34282018

Sanchez AN, Alitto HJ, Usrey WM (2020) Eye to brain: Parallel visual pathways. In: The senses, 2nd edition (Martin PR, ed), pp 362–368. New York: Elsevier.

Sanchez AN, Alitto HJ, Usrey, WM (2021) Center-surround spatial organization of corticogeniculate feed-back. Soc Neurosci Abstr. (in press)

Sanchez-Gonzalez MA, Garcia-Cabezas MA, Rico B, Cavada C (2005) The primate thalamus is a key target for brain dopamine. J Neurosci 25:6076–6083.

Sanchez-Vives MV, Nowak LG, McCormick DA (2000a) Cellular mechanisms of long-lasting adaptation in visual cortical neurons *in vitro*. J Neurosci 20:4286–4299.

Sanchez-Vives MV, Nowak LG, McCormick DA (2000b) Membrane mechanisms underlying contrast adaptation in cat area 17 *in vivo*. J Neurosci 20:4267–4285.

Saul AB, Humphrey AL (1990) Spatial and temporal response properties of lagged and nonlagged cells in cat lateral geniculate nucleus. J Neurophysiol 64:206–224.

Sawai H, Fukuda Y, Wakakuwa K (1985) Axonal projections of X-cells to the superior colliculus and to the nucleus of the optic tract in cats. Brain Res 341:1–6.

Scannell JW, Blakemore C, Young MP (1995) Analysis of connectivity in the cat cerebral cortex. J Neurosci 15:1463–1483.

Schiller PH, Logothetis NK (1990) The color-opponent and broad-band channels of the primate visual system. Trends Neurosci 13:392–398.

Schiller PH, Malpeli JG (1977) Properties and tectal projections of monkey retinal ganglion cells. J Neurophysiol 40:428–445.

Schiller PH, Malpeli JG (1978) Functional specificity of lateral geniculate nucleus laminae of the rhesus monkey. J Neurophysiol 41:788–797.

Schiller PH, Tehovnik EJ (2001) Look and see: How the brain moves your eyes about. Prog Brain Res 134:127–142.

Schmielau F, Singer W (1977) The role of visual cortex for binocular interactions in the cat lateral geniculate nucleus. Brain Res 120:354–361.

Schmitt LI, Wimmer RD, Nakajima M, Happ M, Mofakham S, Halassa MM (2017) Thalamic amplification of cortical connectivity sustains attentional control. Nature 545:219–223.

Schneider KA, Kastner S (2009) Effects of sustained spatial attention in the human lateral geniculate nucleus and superior colliculus. J Neurosci 29:1784–1795.

Scholl B, Tan AY, Corey J, Priebe NJ (2013) Emergence of orientation selectivity in the mammalian visual pathway. J Neurosci 33:10616–10624.

Sclar G, Lennie P, DePriest DD (1989) Contrast adaptation in striate cortex of macaque. Vision Res 29:747–755.

Seabrook TA, Burbridge TJ, Crair MC, Huberman AD (2017) Architecture, function, and assembly of the mouse visual system. Annu Rev Neurosci 40:499–538.

Sedigh-Sarvestani M, Vigeland L, Fernandez-Lamo I, Taylor M, Palmer LA, Contreras D (2017) Intracellular, in vivo, dynamics of thalamocortical synapses in visual cortex. J Neurosci. 37:5250–5262.

Segev I (1998) Cable and compartmental models of dendritic trees. In: The book of GENESIS: Exploring Realistic Neural Models with the GEneral NEural SImulation System (Bower JM, Beeman D, eds), pp 51–77. New York: Springer-Verlag.

Seidemann E, Zohary E, Newsome WT (1998) Temporal gating of neural signals during performance of a visual discrimination task. Nature 394:72–75.

Shadlen MN, Movshon JA (1999) Synchrony unbound: A critical evaluation of the temporal binding hypothesis. Neuron 24:67–77.

Shapley R, Perry VH (1986) Cat and monkey retinal ganglion cells and their visual functional roles. Trends Neurosci 9:229–235.

Shapley RM (1992) Parallel retinocortical channels: X and Y and P and M. In: Applications of parallel processing in vision (Brannan J, ed), pp 3–36. New York: Elsevier.

Sherman SM (1985) Functional organization of the W-,X-, and Y-cell pathways in the cat: A review and hypothesis. In: Progress in psychobiology and physiological psychology, vol. 11 (Sprague JM, Epstein AN, eds), pp 233–314. Orlando, FL: Academic Press.

Sherman SM (1996) Dual response modes in lateral geniculate neurons: Mechanisms and functions. Visual Neurosci 13:205–213.

Sherman SM (2001) Tonic and burst firing: Dual modes of thalamocortical relay. Trends Neurosci 24:122–126.

Sherman SM (2004) Interneurons and triadic circuitry of the thalamus. Trends Neurosci 27:670–675.

Sherman SM (2007) The thalamus is more than just a relay. Curr Opin Neurobiol 17:1–6.

Sherman SM (2013) Information processing in thalamocortical circuits. In: Encyclopedia of pain (Gebhart G, Schmidt R, eds), pp 1–6. Berlin Heidelberg: Springer-Verlag.

Sherman SM (2014) The function of metabotropic glutamate receptors in thalamus and cortex. Neuroscientist 20:136–149.

Sherman SM (2016) Thalamus plays a central role in ongoing cortical functioning. Nat Neurosci 19:533–541.

Sherman SM, Friedlander MJ (1988) Identification of X versus Y properties for interneurons in the A-laminae of the cat's lateral geniculate nucleus. Exp Brain Res 73:384–392.

Sherman SM, Guillery RW (1996) The functional organization of thalamocortical relays. J Neurophysiol 76:1367–1395.

Sherman SM, Guillery RW (1998) On the actions that one nerve cell can have on another: Distinguishing "drivers" from "modulators." Proc Natl Acad Sci USA 95:7121–7126.

Sherman SM, Guillery RW (2001) Exploring the thalamus. San Diego: Academic Press.

Sherman SM, Guillery RW (2006) Exploring the thalamus and its role in cortical function. Cambridge, MA: MIT Press.

Sherman SM, Guillery RW (2013) Functional connections of cortical areas: A new view from the thalamus. Cambridge, MA: MIT Press.

Sherman SM, Guillery RW (2014) The lateral geniculate nucleus and pulvinar. In: The new visual neurosciences (Chalupa LM, Werner JS, eds), pp 257–283. Cambridge, MA: MIT Press.

Sherman SM, Koch C (1986) The control of retinogeniculate transmission in the mammalian lateral geniculate nucleus. Exp Brain Res 63:1–20.

Sherman SM, Spear PD (1982) Organization of visual pathways in normal and visually deprived cats. Physiol Rev 62:738–855.

Sherman SM, Usrey WM (2021) Cortical control of behavior and attention from an evolutionary perspective. Neuron available from: PM:34297915.

Sherman SM, Wilson JR, Kaas JH, Webb SV (1976) X- and Y-cells in the dorsal lateral geniculate nucleus of the owl monkey (*Aotus trivirgatus*). Science 192:475–477.

Sillito AM, Jones HE, Gerstein GL, West DC (1994) Feature-linked synchronization of thalamic relay cell firing induced by feedback from the visual cortex. Nature 369:479–482.

Sincich LC, Adams DL, Economides JR, Horton JC (2007) Transmission of spike trains at the retinogeniculate synapse. J Neurosci 27:2683–2692.

Singer W, Gray CM (1995) Visual feature integration and the temporal correlation hypothesis. Annu Rev Neurosci 18:555–586.

Sitnikova EY, Raevskii VV (2010) The lemniscal and paralemniscal pathways of the trigeminal system in rodents are integrated at the level of the somatosensory cortex. Neurosci Behav Physiol 40:325–331.

Slater BJ, Willis AM, Llano DA (2013) Evidence for layer-specific differences in auditory corticocollicular neurons. Neuroscience 229:144–154.

Smith GD, Cox CL, Sherman SM, Rinzel J (2000) Fourier analysis of sinusoidally-driven thalamocortical relay neurons and a minimal integrate-and-fire-or-burst model. J Neurophysiol 83:588–610.

Smith GD, Sherman SM (2002) Detectability of excitatory versus inhibitory drive in an integrate-and-fire-or-burst thalamocortical relay neuron model. J Neurosci 22:10242–10250.

Smith Y, Séguéla P, Parent A (1987) Distribution of GABA-immunoreactive neurons in the thalamus of the squirrel monkey (*Saimiri sciureus*). Neurosci 22:579–591.

Snow JC, Allen HA, Rafal RD, Humphreys GW (2009) Impaired attentional selection following lesions to human pulvinar: Evidence for homology between human and monkey. Proc Natl Acad Sci USA 106(10):4054–4059.

So YT, Shapley R (1979) Spatial properties of X and Y cells in the lateral geniculate nucleus of the cat and conduction velocities of their inputs. Exp Brain Res 36:533–550.

Solomon SG, Lennie P (2007) The machinery of colour vision. Nat Rev Neurosci 8:276–286.

Solomon SG, Peirce JW, Dhruv NT, Lennie P (2004) Profound contrast adaptation early in the visual pathway. Neuron 42:155–162.

Solomon SG, White AJ, Martin PR (1999) Temporal contrast sensitivity in the lateral geniculate nucleus of a New World monkey, the marmoset Callithrix jacchus. J Physiol (Lond) 517(Pt 3):907–917.

Sommer MA, Wurtz RH (2004a) What the brain stem tells the frontal cortex. I. Oculomotor signals sent from superior colliculus to frontal eye field via mediodorsal thalamus. J Neurophysiol 91:1381–1402.

Sommer MA, Wurtz RH (2004b) What the brain stem tells the frontal cortex. II. Role of the SC-MD-FEF pathway in corollary discharge. J Neurophysiol 91:1403–1423.

Sommer MA, Wurtz RH (2008) Brain circuits for the internal monitoring of movements. Annu Rev Neurosci 31:317–338.

Sottile SY, Hackett TA, Cai R, Ling L, Llano DA, Caspary DM (2017) Presynaptic neuronal nicotinic receptors differentially shape select inputs to auditory thalamus and are negatively impacted by aging. J Neurosci 37(47):11377–11389.

Sperry RW (1950) Neural basis of the spontaneous optokinetic response produced by visual inversion. J Comp Neurol 43:482–489.

Spruston N, Hausser M, Stuart G (2014) Information processing in dendrites and spines. In: Fundamental neuroscience, fourth edition (Squire LR, Berg D, Bloom FE, du Lac S, Ghosh A, Spitzer N, eds), pp 231–260. Oxford: Academic Press.

Stanford LR, Friedlander MJ, Sherman SM (1981) Morphology of physiologically identified W-cells in the C laminae of the cat's lateral geniculate nucleus. J Neurosci 1:578–584.

Stanford LR, Friedlander MJ, Sherman SM (1983) Morphological and physiological properties of geniculate W-cells of the cat: A comparison with X- and Y-cells. J Neurophysiol 50:582–608.

Stein BE, Gaither NS (1983) Receptive-field properties in reptilian optic tectum: Some comparisons with mammals. J Neurophysiol 50:102–124.

Stein BE, Stanford TR, Rowland BA (2009) The neural basis of multisensory integration in the midbrain: Its organization and maturation. Hear Res 258:4–15.

Stepniewska I, Qi HX, Kaas JH (1999) Do superior colliculus projection zones in the inferior pulvinar project to MT in primates? Eur J Neurosci 11:469–480.

Stepniewska I, Qi HX, Kaas JH (2000) Projections of the superior colliculus to subdivisions of the inferior pulvinar in New World and Old World monkeys. Visual Neurosci 17:529–549.

Steriade M, McCarley RW (1990) Brainstem control of wakefulness and sleep. New York: Plenum Press.

Steriade M, McCarley RW (2005) Brain control of sleep and wakefulness. New York: Kluwer Academic Publishers.

Stoelzel CR, Bereshpolova Y, Swadlow HA (2009) Stability of thalamocortical synaptic transmission across awake brain states. J Neurosci 29:6851–6859.

Stone J (1983) Parallel processing in the visual system. New York: Plenum Press.

Stratford KJ, Tarczy-Hornoch K, Martin KAC, Bannister NJ, Jack JJB (1996) Excitatory synaptic inputs to spiny stellate cells in cat visual cortex. Nature 382:258–261.

Stroh A, Adelsberger H, Groh A, Ruhlmann C, Fischer S, Schierloh A, Deisseroth K, Konnerth A (2013) Making waves: Initiation and propagation of corticothalamic ca(2+) waves in vivo. Neuron 77:1136–1150.

Stryker MP, Zahs KR (1983) On and off sublaminae in the lateral geniculate nucleus of the ferret. J Neurosci 3:1943–1951.

Stuart G, Spruston N, Sakmann B, Hausser M (1997) Action potential initiation and backpropagation in neurons of the mammalian CNS. Trends Neurosci 20:125–131.

Suga N (2018) Specialization of the auditory system for the processing of bio-sonar information in the frequency domain: Mustached bats. Hear Res 361:1–22.

Sur M, Esguerra M, Garraghty PE, Kritzer MF, Sherman SM (1987) Morphology of physiologically identified retinogeniculate X- and Y-axons in the cat. J Neurophysiol 58:1–32.

Sur M, Sherman SM (1982) Linear and nonlinear W-cells in C-laminae of the cat's lateral geniculate nucleus. J Neurophysiol 47:869–884.

Suzuki DG, Perez-Fernandez J, Wibble T, Kardamakis AA, Grillner S (2019) The role of the optic tectum for visually evoked orienting and evasive movements. Proc Natl Acad Sci USA 116:15272–15281.

Suzuki M, Larkum ME (2020) General anesthesia decouples cortical pyramidal neurons. Cell 180(4):666–676.

Swadlow HA, Gusev AG (2001) The impact of "bursting" thalamic impulses at a neocortical synapse. Nat Neurosci 4:402–408.

Swadlow HA, Gusev AG, Bezdudnaya T (2002) Activation of a cortical column by a thalamocortical impulse. J Neurosci 22:7766–7773.

Szel A, Rohlich P (1992) Two cone types of rat retina detected by anti-visual pigment antibodies. Exp Eye Res 55(1):47–52.

Tamamaki N, Uhlrich DJ, Sherman SM (1994) Morphology of physiologically identified retinal X and Y axons in the cat's thalamus and midbrain as revealed by intra-axonal injection of biocytin. J Comp Neurol 354:583–607.

Tamamaki N, Uhlrich DJ, Sherman SM (1995) Morphology of physiologically identified retinal X and Y axons in the cat's thalamus and midbrain as revealed by intra-axonal injection of biocytin. J Comp Neurol 354:583–607.

Tan Y, Li WH (1999) Vision: Trichromatic vision in prosimians. Nature 402:36.

Tanaka K (1985) Organization of geniculate, inputs to visual cortical cells in the cat. Vision Res 25:357–364.

Tanibuchi I, Kitano H, Jinnai K (2009) Substantia nigra output to prefrontal cortex via thalamus in monkeys. I. Electrophysiological identification of thalamic relay neurons. J Neurophysiol 102(5):2933–2945.

Tehovnik EJ, Slocum WM, Schiller PH (2002) Differential effects of laminar stimulation of V1 cortex on target selection by macaque monkeys. Eur J Neurosci 16:751–760.

Tham WW, Stevenson RJ, Miller LA (2009) The functional role of the medio dorsal thalamic nucleus in olfaction. Brain Res Rev 62:109–126.

Theyel BB, Llano DA, Sherman SM (2010) The corticothalamocortical circuit drives higher-order cortex in the mouse. Nat Neurosci 13:84–88.

Thompson SM, Capogna M, Scanziani M (1993) Presynaptic inhibition in the hippocampus. Trends Neurosci 16:222–227.

Thomson AM, Deuchars J (1994) Temporal and spatial properties of local circuits in neocortex. Trends Neurosci 17:119–126.

Thomson AM, Deuchars J (1997) Synaptic interactions in neocortical local circuits: Dual intracellular recordings *in vitro*. Cereb Cortex 7:510–522.

Tootell RBH, Silverman MS, Switkes E, De Valois RL (1982) Deoxyglucose analysis of retinotopic organizaiton in primate striate cortex. Science 218:902–904.

Troy JB, Lennie P (1987) Detection latencies of X and Y type cells of the cat's dorsal lateral geniculate nucleus. Exp Brain Res 65:703–706.

Tsanov M, Manahan-Vaughan D (2009) Synaptic plasticity in the adult visual cortex is regulated by the metabotropic glutamate receptor, mGLUR5. Exp Brain Res 199:391–399.

Tsumoto T, Creutzfeldt OD, Legendy CR (1978) Functional organization of the cortifugal system from visual cortex to lateral geniculate nucleus in the cat. Exp Brain Res 32:345–364.

Tsumoto T, Suda K (1980) Three groups of cortico-geniculate neurons and their distribution in binocular and monocular segments of cat striate cortex. J Comp Neurol 193:223–236.

Tu S, Menke RAL, Talbot K, Kiernan MC, Turner MR (2018) Regional thalamic MRI as a marker of widespread cortical pathology and progressive frontotemporal involvement in amyotrophic lateral sclerosis. J Neurol Neurosurg Psychiatry 89(12):1250–1258.

Tuncdemir SN, Wamsley B, Stam FJ, Osakada F, Goulding M, Callaway EM, Rudy B, Fishell G (2016) Early somatostatin interneuron connectivity mediates the maturation of deep layer cortical circuits. Neuron 89(3):521–535.

Uhlrich DJ, Cucchiaro JB, Humphrey AL, Sherman SM (1991) Morphology and axonal projection patterns of individual neurons in the cat perigeniculate nucleus. J Neurophysiol 65:1528–1541.

Umans JG, Levi R (1995) Nitric oxide in the regulation of blood flow and arterial pressure. Annu Rev Physiol 57:771–790.

Usrey WM (2002a) Spike timing and visual processing in the retinogeniculocortical pathway. Philos Trans R Soc Lond [Biol] 357:1729–1737.

Usrey WM (2002b) The role of spike timing for thalamocortical processing. Curr Opin Neurobiol 12:411–417.

Usrey WM, Alitto HJ (2015) Visual functions of the thalamus. Annu Rev Vis Sci 1:351–371.

Usrey WM, Alonso JM, Reid RC (2000) Synaptic interactions between thalamic inputs to simple cells in cat visual cortex. J Neurosci 20:5461–5467.

Usrey WM, Fitzpatrick D (1996) Specificity in the axonal connections of layer VI neurons in tree shrew striate cortex: Evidence for distinct granular and supragranular systems. J Neurosci 16:1203–1218.

Usrey WM, Kastner S (2020) Functions of the visual thalamus in selective attention. In: The cognitive neurosciences, 6th edition (Poeppel D, Gazzaniga MS, eds), pp 369–377. Cambridge, MA: MIT Press.

Usrey WM, Muly EC, Fitzpatrick D (1992) Lateral geniculate projections to the superficial layers of visual cortex in the tree shrew. J Comp Neurol 319:159–171.

Usrey WM, Reid RC (1999) Synchronous activity in the visual system. Ann Rev Physiol 61:435–456.

Usrey WM, Reid RC (2000) Visual physiology of the lateral geniculate nucleus in two species of New World monkey: Saimiri sciureus and Aotus trivirgatis. J Physiol 523:755–769.

Usrey WM, Reppas JB, Reid RC (1998) Paired-spike interactions and synaptic efficacy of retinal inputs to the thalamus. Nature 395:384–387.

Usrey WM, Reppas JB, Reid RC (1999) Specificity and strength of retinogeniculate connections. J Neurophysiol 82:3527–3540.

Usrey WM, Sherman SM (2019) Corticofugal circuits: Communication lines from the cortex to the rest of the brain. J Comp Neurol 527:640–650.

Van den Pol AN (2012) Neuropeptide transmission in brain circuits. Neuron 76(1):98–115.

Van Essen DC, Anderson CH, Felleman DJ (1992) Information processing in the primate visual system: An integrated systems perspective. Science 255:419–423.

Van Horn SC, Erisir A, Sherman SM (2000) The relative distribution of synapses in the A-laminae of the lateral geniculate nucleus of the cat. J Comp Neurol 416:509–520.

Van Horn SC, Sherman SM (2004) Differences in projection patterns between large and small corticothalamic terminals. J Comp Neurol 475:406–415.

Van Horn SC, Sherman SM (2007) Fewer driver synapses in higher order than in first order thalamic relays. Neurosci 475:406–415.

Vaney DI, Peichl L, Wässle H, Illing R-B (1981) Almost all ganglion cells in the rabbit retina project to the superior colliculus. Brain Res 212:447–453.

Vantomme G, Osorio-Forero A, Luthi A, Fernandez LMJ (2019) Regulation of local sleep by the thalamic reticular nucleus. Front Neurosci 13:576.

Varela C, Sherman SM (2007) Differences in response to muscarinic agonists between first and higher order thalamic relays. J Neurophysiol 98:3538–3547.

Varela C, Sherman SM (2008) Differences in response to serotonergic activation between first and higher order thalamic nuclei. Cereb Cortex 19:1776–1786.

Veinante P, Lavallée P, Deschênes M (2000) Corticothalamic projections from layer 5 of the vibrissal barrel cortex in the rat. J Comp Neurol 424:197–204.

Venkatadri PS, Lee CC (2014) Differential expression of mGluR2 in the developing cerebral cortex of the mouse. J Biomed Sci Eng 7:1030–1037.

Viaene AN, Petrof I, Sherman SM (2011a) Properties of the thalamic projection from the posterior medial nucleus to primary and secondary somatosensory cortices in the mouse. Proc Nat Acad Sci USA 108:18156–18161.

Viaene AN, Petrof I, Sherman SM (2011b) Synaptic properties of thalamic input to layers 2/3 in primary somatosensory and auditory cortices. J Neurophysiol 105:279–292.

Viaene AN, Petrof I, Sherman SM (2011c) Synaptic properties of thalamic input to the subgranular layers of primary somatosensory and auditory cortices in the mouse. J Neurosci 31:12738–12747.

Viaene AN, Petrof I, Sherman SM (2013) Activation requirements for metabotropic glutamate receptors. Neurosci Lett 541:67–72.

Vitten H, Isaacson JS (2001) Synaptic transmission: Exciting times for presynaptic receptors. Curr Biol 11:R695–R697.

von Holst E, Mittelstaedt H (1950) The reafference principle. Interaction between the central nervous system and the periphery. In: Selected papers of Erich von Holst: The behavioural physiology of animals and man (translated by Robert Martin, ed), pp 139–173. Coral Gables, FL: University of Miami Press.

Wang S, Bickford ME, Van Horn SC, Erisir A, Godwin DW, Sherman SM (2001) Synaptic targets of thalamic reticular nucleus terminals in the visual thalamus of the cat. J Comp Neurol 440:321–341.

Wang S, Eisenback MA, Bickford ME (2002) Relative distribution of synapses in the pulvinar nucleus of the cat: Implications regarding the "driver/modulator" theory of thalamic function. J Comp Neurol 454:482–494.

Wang W, Andolina IM, Lu Y, Jones HE, Sillito AM (2018) Focal gain control of thalamic visual receptive fields by layer 6 corticothalamic feedback. Cereb Cortex 28(1):267–280.

Wang X, Hirsch JA, Sommer FT (2010) Recoding of sensory information across the retinothalamic synapse. J Neurosci 30:13567–13577.

Wang W, Jones HE, Andolina IM, Salt TE, Sillito AM (2006) Functional alignment of feedback effects from visual cortex to thalamus. Nat Neurosci 9:1330–1336.

Wang X, Sommer FT, Hirsch JA (2011) Inhibitory circuits for visual processing in thalamus. Curr Opin Neurobiol 21:726–733.

Wässle H (1982) Morphological types and central projections of ganglion cells in the cat retina. In: Progress in retinal research (Osborne N, Chader G, eds), pp 125–152. New York: Pergamon Press.

Wässle H (2004) Parallel processing in the mammalian retina. Nat Rev Neurosci 5:747–757.

Wässle H, Grünert U, Röhrenbeck J, Boycott BB (1989) Cortical magnification factor and the ganglion cell density of the primate retina. Nature 341:643–646.

Wässle H, Illing R-B (1980) The retinal projection to the superior colliculus in the cat: A quantitative study with HRP. J Comp Neurol 190:333–356.

Watanabe M, Rodieck RW (1989) Parasol and midget ganglion cells of the primate retina. J Comp Neurol 289:434–454.

Weber JT, Huerta MF, Kaas JH, Harting JK (1983) The projections of the lateral geniculate nucleus of the squirrel monkey: Studies of the interlaminar zones and the S layers. J Comp Neurol 213(2):135–145.

Wei H, Masterson SP, Petry HM, Bickford ME (2011) Diffuse and specific tectopulvinar terminals in the tree shrew: Synapses, synapsins, and synaptic potentials. PLoS One 6:e23781.

Weliky M (1999) Recording and manipulating the *in vivo* correlational structure of neuronal activity during visual cortical development. J Neurobiol 41:25–32.

Wernet MF, Huberman AD, Desplan C (2014) So many pieces, one puzzle: Cell type specification and visual circuitry in flies and mice. Genes Dev 28(23):2565–2584.

Weyand TG (2007) Retinogeniculate transmission in wakefulness. J Neurophysiol 98:769–785.

White AJ, Wilder HD, Goodchild AK, Sefton AJ, Martin PR (1998) Segregation of receptive field properties in the lateral geniculate nucleus of a new-world monkey, the marmoset *Callithrix jacchus*. J Neurophysiol 80:2063–2076.

White AJR, Solomon SG, Martin PR (2001) Spatial properties of koniocellular cells in the lateral geniculate nucleus of the marmoset *Callithrix jacchus*. J Physiol 533:519–535.

Wilke M, Turchi J, Smith K, Mishkin M, Leopold DA (2010) Pulvinar inactivation disrupts selection of movement plans. J Neurosci 30(25):8650–8659.

Williams LE, Holtmaat A (2019) Higher-order thalamocortical inputs gate synaptic long-term potentiation via disinhibition. Neuron 101(1):91–102.

Wilson JR, Friedlander MJ, Sherman SM (1984) Fine structural morphology of identified X- and Y-cells in the cat's lateral geniculate nucleus. Proc Roy Soc Lond B 221:411–436.

Wilson JR, Sherman SM (1976) Receptive-feild characteristics of neurons in cat striate cortex: Changes with visual field eccentricity. J Neurophysiol 39:512–533.

Wilson PD, Rowe MH, Stone J (1976) Properties of relay cells in cat's lateral geniculate nucleus: A comparison of W-cells with X- and Y-cells. J Neurophysiol 39:1193–1209.

Wimmer RD, Schmitt LI, Davidson TJ, Nakajima M, Deisseroth K, Halassa MM (2015) Thalamic control of sensory selection in divided attention. Nature 526:705–709.

Winer JA, Miller LM, Lee CC, Schreiner CE (2005) Auditory thalamocortical transformation: Structure and function. Trends Neurosci 28:255–263.

Winnubst J, et al. (2019) Reconstruction of 1,000 projection neurons reveals new cell types and organization of long-range connectivity in the mouse brain. Cell 179:268–281.

Wolpert DM, Flanagan JR (2010) Motor learning. Curr Biol 20:R467–R472.

Womelsdorf T, Fries P, Mitra PP, Desimone R (2006) Gamma-band synchronization in visual cortex predicts speed of change detection. Nature 439:733–736.

Wu LG, Saggau P (1997) Presynaptic inhibition of elicited neurotransmitter release. Trends Neurosci 20:204–212.

Wurtz RH (2009) Recounting the impact of Hubel and Wiesel. J Physiol 587(Pt 12):2817–2823.

Wurtz RH, Sommer MA (2004) Identifying corollary discharges for movement in the primate brain. Prog Brain Res 144:47–60.

Xiao D, Zikopoulos B, Barbas H (2009) Laminar and modular organization of prefrontal projections to multiple thalamic nuclei. Neurosci 161:1067–1081.

Yantis S, Schwarzbach J, Serences JT, Carlson RL, Steinmetz MA, Pekar JJ, Courtney SM (2002) Transient neural activity in human parietal cortex during spatial attention shifts. Nat Neurosci 5(10):995–1002.

Yen C-T, Jones EG (1983) Intracellular staining of physiologically identified neurons and axons in the somatosensory thalamus of the cat. Brain Res 280:148–154.

Yu C, Derdikman D, Haidarliu S, Ahissar E (2006) Parallel thalamic pathways for whisking and touch signals in the rat. PloS Biol 4:e124.

Zhan XJ, Cox CL, Rinzel J, Sherman SM (1999) Current clamp and modeling studies of low threshold calcium spikes in cells of the cat's lateral geniculate nucleus. J Neurophysiol 81:2360–2373.

Zhan XJ, Cox CL, Sherman SM (2000) Dendritic depolarization efficiently attenuates low threshold calcium spikes in thalamic relay cells. J Neurosci 20:3909–3914.

Zhang J, Snyder SH (1995) Nitric oxide in the nervous system. Annu Rev Pharmacol Toxicol 35:213–233.

Zhou H, Schafer RJ, Desimone R (2016) Pulvinar-cortex interactions in vision and attention. Neuron 89:209–220.

Zhou NA, Maire PS, Masterson SP, Bickford ME (2017) The mouse pulvinar nucleus: Organization of the tectorecipient zones. Vis Neurosci 34:E011.

Zwergal A, la FC, Lorenzl S, Rominger A, Xiong G, Deutschenbaur L, Linn J, Krafczyk S, Dieterich M, Brandt T, Strupp M, Bartenstein P, Jahn K (2011) Postural imbalance and falls in PSP correlate with functional pathology of the thalamus. Neurology 77(2):101–109.

Index

Tables and figures are indicated by *t* and *f* following the page number.